Lecture Notes in Mathematics 1506

Alexandru Buium

Differential Algebraic Groups of Finite Dimension

Springer-Verlag
Berlin Heidelberg New York
London Paris Tokyo
Hong Kong Barcelona
Budapest

Author

Alexandru Buium
Institute of Mathematics
of the Romanian Academy
P. O. Box 1–764
RO-70700 Bucharest, Romania

Mathematics Subject Classification (1991): 12H05, 14L17, 14K10, 14G05, 20F28

ISBN 3-540-55181-6 Springer-Verlag Berlin Heidelberg New York
ISBN 0-387-55181-6 Springer-Verlag New York Berlin Heidelberg

Printed in Germany

Typesetting: Camera ready by author
Printing and binding: Druckhaus Beltz, Hemsbach/Bergstr.
46/3140-543210 - Printed on acid-free paper

INTRODUCTION

1. Scope

Differential algebraic groups are defined roughly speaking as "groups of solutions of algebraic differential equations" in the same way in which algebraic groups are defined as "groups of solutions of algebraic equations". They were introduced in modern literature by Cassidy [C_1] and Kolchin [K_2]; their pre-history goes back however to classical work of S. Lie, E. Cartan and J.F. Ritt.

Let's contemplate a few examples before giving the formal definition (for which we send to Section 3 of this Introduction). Start with any linear differential equation:

$$y^{(n)} + a_1 y^{(n-1)} + \dots + a_n y = 0 \tag{1}$$

where the unknown y and the coefficients a_i are, say, meromorphic functions of the complex variable t. The difference of any two solutions of (1) is again a solution of (1); so the solutions of (1) form a group with respect to addition. This provides a first example of "differential algebraic group".

Similarily, consider the system

$$\begin{cases} xy - 1 = 0 \\ yy'' - (y')^2 + ayy' = 0 \end{cases} \tag{2}$$

where the unknowns x,y and the coefficient a are once again meromorphic in t. This system (extracted from a paper of Cassidy [C_4]) has the property that the quotient $(x_1/x_2, y_1/y_2)$ of any two solutions (x_1,y_1), (x_2,y_2) is again a solution; so the solutions of (2) form a group with respect to the multiplicative group of the hyperbola $xy - 1 = 0$ and we are led to another example of "differential algebraic group".

Examples of a more subtle nature are provided by the systems

$$\begin{cases} y^2 - x(x-1)(x-c) = 0 \\ x''y - x'y' + ax'y = 0 \end{cases} \tag{3a}$$

$$\begin{cases} y^2 - x(x-1)(x-t) = 0 \\ -y^3 - 2(2t-1)(x-t)^2x'y + 2t(t-1)(x-t)^2(x''y - 2x'y') = 0 \end{cases} \tag{3b}$$

where the unknowns x, y are still meromorphic functions in t, the coefficient a is meromorphic in t and $c \neq 0,1$ is a constant in $\mathbb{C}$. These systems (of which (3b) is extracted from a paper of Manin [Ma]) have the property that if (x_1,y_1) and (x_2,y_2) are distinct solutions of one of the systems, then their difference, in the group law of the elliptic curve A defined by the first equation of that system, is again a solution; so the solutions of (3a), (3b) together with the point at infinity (0 : 1 : 0) of A form groups with respect to the group law of A and lead to other examples of "differential algebraic groups".

A differential algebraic group will be called of finite dimension if roughly speaking its elements "depend on finitely many integration constants" rather than on "arbitrary functions". This is the case with the "differential algebraic groups" derived from (1), (2), (3); on the contrary, for instance, the group, say, of all matrices

$$\begin{pmatrix} x & 0 & y \\ 0 & 1 & z \\ 0 & 0 & 1 \end{pmatrix}, \qquad x \neq 0$$

satisfying the system

$$(4) \qquad \begin{cases} z' = 0 \\ x' = zx \end{cases}$$

has its elements depending on the "arbitrary function" y so it is an "infinite dimensional differential algebraic group". Differential algebraic groups are known today but to a few mathematicians; and this is because Kolchin's language [K_1][K_2] through which they are studied is known but to a few mathematicians. Yet differential algebraic groups certainly deserve a much broader audience, especially among algebraic geometers. The scope of the present research monograph is twofold namely: 1) to provide an algebraic geometer's introduction to differential algebraic groups of finite dimension and 2) to develop a structure and classification theory for these groups. Unless otherwise stated, all results appearing in this book are due to the author and were never published before.

2. Motivation

The original motivation for the study of differential algebraic groups in [C_i] and [K_2] was undoubtedly their intrinsic beauty and variety. Admittedly no such group appeared so far to play a role, say, in mathematical physics; but as we hope to demonstrate in the present work, differential algebraic groups play a role in algebraic geometry. This is already suggested for instance by the implicit occurence of the differential algebraic group (3b) in Manin's paper [Ma] on the geometric Mordell conjecture. Moreover as shown in the author's papers [B_6], [B_7] differential algebraic groups may be used along a line quite different from Manin's to prove the geometric and infinitesimal analogues of a diophantine conjecture of S. Lang (about intersections of subvarieties of abelian varieties with finite rank subgroups). The latter application will be presented without proof in an appendix. Other applications and interesting links with other topics of algebraic geometry (such as deformations of algebraic groups and their automorphisms, moduli spaces of abelian varieties, the Grothendieck-Mazur-Messing crystalline theory of universal extensions of abelian varieties [MM], a.s.o.) will appear in the body of the text (and will also be touched in this Introduction).

3. Formalization

Before explaining our strategy and results in some detail it will be convenient to provide the formal definition of differential algebraic groups with which we will operate in this book.

The frame in which this definition will be given is that of "differential algebraic geometry" by which we mean here the analogue (due to Ritt and Kolchin) of algebraic geometry in which algebraic equations are replaced by algebraic differential equations. Let's quickly review the basic concepts of this geometry, cf. [R], [K_1], [C_1]; for details we send to the first section of Chapter 5 of this book.

One starts with a field of characteristic zero $\mathcal{U}$ equiped with a derivation $\delta : \mathcal{U} \to \mathcal{U}$ (i.e. with an additive map satisfying $\delta(xy) = (\delta x)y + x(\delta y)$ for all $x,y \in \mathcal{U}$) and set $\mathcal{K} = \{x \in \mathcal{U};\ \delta x = 0\}$ the field of constants. We assume that $\mathcal{U}$ is "sufficiently big" as to "contain" all "solutions of algebraic differential equations with coefficients in it"; for the reader familiar with [K_1] what we assume is that $\mathcal{U}$ is "universal". Such a field $\mathcal{U}$ is a quite artificial being; but since all problems related to algebraic differential equations can be "embedded" into "problems over $\mathcal{U}$" the use of this field appears to be extremely useful and in any case it greatly simplifies language. Next one considers the ring of differential polynomials (shortly Δ - polynomials) $\mathcal{U}\{y_1, \dots y_n\}$ which by definition is the ring of polynomials with coefficients in $\mathcal{U}$ in the indeterminates

$$y_1, \dots y_n, y'_1, \dots, y'_n, \dots, y_1^{(k)}, \dots, y_n^{(k)}, \dots$$

E.g. the expressions appearing in the left hand side of the equations (1), (2), (3) (where we assume now $a_i, a, t \in \mathcal{U}$, $t' = 1$, $c \in \mathcal{K}$) are Δ - polynomials. Now for any finite set $F_1, \dots, F_m$ of Δ - polynomials we may consider their set Σ of common zeroes in the affine space $\mathbf{A}^n(\mathcal{U}) = \mathcal{U}^n$; such a set will be called a Δ - closed subset of $\mathcal{U}^n$. Δ - closed sets form a Noetherian topology on $\mathcal{U}^n$ called the Δ - topology which is stronger than the Zariski topology. For instance the equation (1) defines an irreducible Δ - closed subset of the affine line $\mathbf{A}^1(\mathcal{U}) = \mathcal{U}$ while the systems (2) and (3) define irreducible Δ - closed subset of the affine plane $\mathbf{A}^2(\mathcal{U}) = \mathcal{U}^2$.

Any irreducible Δ - closed subset Σ in $\mathcal{U}^n$ has a natural sheaf (for the induced Δ - topology) of $\mathcal{U}$- valued functions on it called the sheaf of Δ - regular functions; roughly speaking a function on a Δ - open set of Σ is called Δ - regular if, locally in the Δ - topology, it is given by a quotient of Δ - polynomials in the coordinates. Σ together with this sheaf will be called an affine differential algebraic manifold (shortly, an affine Δ - manifold).

An irreducible ringed space locally isomorphic to an affine Δ - manifold will be called a differential algebraic manifold (or simply a Δ - manifold). Given a Δ - manifold Σ, the direct limit, over all Δ - open sets Ω of Σ, of the rings of Δ - regular functions on Ω is a field denoted by $\mathcal{U}<\Sigma>$. The transcendence degree of $\mathcal{U}<\Sigma>$ over $\mathcal{U}$ will be called the dimension of Σ and intuitively represents the number of "integration constants" of which the points of Σ depend. E.g. the Δ - closed sets given by equations (1), (2), (3) have dimensions n, 2, 2 respectively.

A map between two Δ - manifolds will be called Δ - regular if it is continuous and pulls back Δ - regular functions into Δ - regular functions. We are provided thus with a category which has direct products called the category of Δ - manifolds. Note that any algebraic $\mathcal{U}$- variety X (respectively any $\mathcal{K}$- variety X_o) has a natural structure of Δ - manifold which we denote by $\hat{X}$ (respectively by $\check{X}_o$); we have $\dim \check{X}_o = \dim X_o$ and $\dim \hat{X} = \infty$ provided $\dim X > 0$.

Finally we define differential algebraic groups (or simply Δ - group) as being group objects in the category of Δ - manifolds, i.e. Δ - manifolds Γ whose set of points is given a group law such that the multiplication $\Gamma \times \Gamma \to \Gamma$ and the inverse $\Gamma \chi \Gamma$ are Δ - regular maps.

Differential algebraic groups of finite dimension will be simply called Δ_o - groups.

The equations (1), (2) clearly provide examples of Δ_o - groups. As for equations (3) the Δ_o - groups which can be derived are not the Δ - closed subsets of $\mathbf{A}^2(\mathcal{U})$ given by (3) but their Δ - closures in $(\mathbf{P}^2)^\wedge$, which contain an additional point (0 : 1 : 0), the origin of these groups.

We leave open the question whether our Δ - groups "are" the same with Kolchin's [K_2]; one can show that any Δ - group in our sense "is" a Δ - group in Kolchin's sense. Moreover one can show that our Δ_o - groups "are" precisely Kolchin's irreducible "Δ - groups of Δ - type zero". Finally note that our concept of "dimension" corresponds to Ritt's concept of "order" [R] and also (in case we have "finite dimension") with Kolchin's concept of "typical Δ - dimension" [K_1][K_2].

4. Strategy

Cassidy and Kolchin developed their theory of differential algebraic groups in analogy with the theory of algebraic group. Our viewpoint will be quite different: we will base our approach on investigating the relations, not the analogies, between the two theories.

Our strategy has two steps. Step 1 will consist in developing a theory of what we call "algebraic D-groups"; this will be done in Chapters 1-4 of the book. Then Step 2 will consist in applying the latter theory to the study of Δ_o - groups; this will be done in Chapter 5.

Let's explain the concept of algebraic D-group; for convenience we will give in this Introduction a rather restricted definition of it (which will be "enlarged" in the body of the text). Let $\mathcal{U}$, δ, $\mathcal{K}$ be as in Section 3 above and let

$$D = \mathcal{U}[\delta] = \sum_{i \geq 0} \mathcal{U}\delta^i \qquad \text{(direct sum)}$$

be the $\mathcal{K}$ - algebra of linear differential operators generated by $\mathcal{U}$ and δ. By an algebraic D-group we will understand an irreducible algebraic $\mathcal{U}$- group G whose structure sheaf $\mathcal{O}_G$ of regular functions is given a structure of sheaf of D - modules such that the multiplication, comultiplication, antipode and co-unit are D-module maps; in other words if $\mu : G \times G \to G$, $S : G \to G$ are the group multiplication and inverse and if $e \in G$ is the unit then for any regular functions ϕ, ψ defined on some open set of G we have the formulae

$$(5)\qquad \begin{aligned} &\delta(\phi\psi) = (\delta\phi)\psi + \phi(\delta\psi) \\ &(\delta\phi)\circ\mu = (\delta\otimes 1 + 1\otimes\delta)(\phi\circ\mu) \\ &\delta(\phi\circ S) = (\delta\phi)\circ S \\ &\delta(\phi(e)) = (\delta\phi)(e) \end{aligned}$$

Some people might like to call such a structure "an algebraic group with connection along δ"; we were inspired in our terminology by the paper of Nichols and Weisfeiler [NW]. Note however that unlike in [NW] we do not assume G is affine, imposing instead that G is of finite type over $\mathcal{U}$! Algebraic D-groups entirely belong to "algebraic geometry" (rather than to the Ritt-Kolchin "differential algebraic geometry") hence Step 1 will inevitably be performed in the field of algebraic geometry; the task of classifying algebraic D-groups will be sometimes quite technical but in the end rewarding.

Step 2 will be based on the result that "the category of Δ_o - groups" is equivalent to "the category of algebraic D - groups"; for any Δ_o - group Γ we shall denote by $G(\Gamma)$ the corresponding algebraic D-group. Note that (the underlying group of) Γ appears as the group of all points $\alpha \in G(\Gamma)(\mathcal{U})$ for which the evaluation map $\mathcal{O}_{G(\Gamma),\alpha} \to \mathcal{U}$ is a D-module map; actually Γ appears as a Δ - closed subgroup of $G(\Gamma)^{\wedge}$. Note that the function field $\mathcal{U}(G(\Gamma))$ identifies with $\mathcal{U}<\Gamma>$. Then Step 2 will deal with the (sometimes not so obvious) translation of properties of algebraic D-groups into properties of Δ_o - groups. In order to get a feeling about the correspondence $\Gamma \mapsto G(\Gamma)$ let's look at the examples we began with (equations (1), (2), (3)).

If Γ is derived from equation (1) then $G(\Gamma)$ is the algebraic vector group $G_a^n = \operatorname{Spec}\mathcal{U}[\xi_o,\xi_1,\dots\xi_{n-1}]$ equiped with the D-module structure of its coordinate algebra defined by

$$\begin{aligned} &\delta\xi_o = \xi_1 \\ &\delta\xi_1 = \xi_2 \\ &\dots\dots\dots \\ &\delta\xi_{n-1} = -a_1\xi_{n-1} - a_2\xi_{n-2} - \dots - a_n\xi_o \end{aligned}$$

(and by the condition that δ is a derivation of $\mathcal{U}[\xi_o,\dots,\xi_{n-1}]$).

If Γ is derived from equation (2) then $G(\Gamma)$ is $G_m \times G_a = \operatorname{Spec}\mathcal{U}[\chi,\chi^{-1},\xi]$, the product of the multiplicative and additive groups, equiped with the D-structure defined by

$$\begin{aligned} &\delta\chi = \chi\xi \\ &\delta\xi = -a\xi \end{aligned}$$

In case Γ is derived from (3a), $G(\Gamma)$ can be proved to be the product $A \times G_a$ while in case Γ is derived from (3b), $G(\Gamma)$ is a non-trivial extension of the elliptic curve A by G_a. Note that in the latter case $G(\Gamma)$ does not descend to $\mathcal{K}$ (because A doesn't).

5. Results

According to the preceeding section the classification problem for Δ_o - groups reduces to answering the following question: given an algebraic $\mathcal{U}$- group G describe all structures of algebraic D-groups on G; call their set P(G/δ). Note that P(G/δ) is a principal homogeneous space for the $\mathcal{U}$- linear space P(G/$\mathcal{U}$) of all $\mathcal{U}$- linear maps $D : \mathcal{O}_G \to \mathcal{O}_G$ satisfying equations (5) with D instead of δ. So we are faced with two problems here namely:

1) What irreducible algebraic $\mathcal{U}$- groups G admit at least one structure of algebraic D - group ?

2) Describe P(G/$\mathcal{U}$) for any such G.

Both these problems have a deformation - theoretic flavour: the first is related to deformations of the algebraic groups themselves while the second is related to infinitesimal deformations of automorphisms of our groups.

Let's consider these problems separately and start by stating our results on the first of them.

THEOREM 1. Let G be an affine algebraic $\mathcal{U}$- group. Then G admits a structure of algebraic D - group if and only if G descends to $\mathcal{K}$.

The proof of Theorem 1 will be done in Chapter 2 and will involve analytic arguments, specifically results of Mostow-Hochschildt [HM_1] and Hamm [Ha]. We already noted that Theorem 1 may fail in the non-affine case; we shall describe in what follows a complete answer to problem 1) in the commutative (non-affine) case. The formulation of the next results requires some familiarity with [Se], [KO]. Their proofs will be done in Chapter 3.

So assume G is an irreducible commutative algebraic $\mathcal{U}$- group and make the following notations:

$L(G)$ = Lie algebra of G

$X_m(G) = \mathrm{Hom}(G,G_m)$ = group of multiplicative characters of G

$X_a(G) = \mathrm{Hom}(G,G_a)$ = group of additive characters of G.

Recall [KO] that we dispose of the Gauss-Manin connection:

$$\nabla : \mathrm{Der}_{\mathcal{K}} \mathcal{U} \to \mathrm{Hom}_{\mathcal{K}}(H^1_{DR}(A), H^1_{DR}(A)), \qquad p \mapsto \nabla_p$$

on the de Rham cohomology space $H^1_{DR}(A)$ of the abelian part A of G (where A = G/B, B = linear part of G = maximum linear connected subgroup of G).

We will introduce in Chapter 3 a "multiplicative analogue" of the Gauss-Manin connection which is a $\mathcal{U}$- linear map

$$\ell\nabla : \mathrm{Der}_{\mathcal{K}} \mathcal{U} \to \mathrm{Hom}_{gr}(H^1_{DR}(A)_m, H^1_{DR}(A)), \qquad p \mapsto \ell\nabla_p$$

where $H^1_{DR}(A)_m$ is the first hypercohomology group of the complex [MM]:

$$1 \to \mathcal{O}^*_A \xrightarrow{\mathrm{dlog}} \Omega^1_{A/\mathcal{U}} \xrightarrow{d} \Omega^2_{A/\mathcal{U}} \xrightarrow{d} \cdots$$

View now G as an extension of A by B, let $S(G)_a$ and $S(G)_m$ be the images of the natural maps $X_a(B) \to H^1(\mathcal{O}_A)$ and $X_m(B) \to \mathrm{Pic}^0(A)$ [Se] and let $S_{DR}(G)_a$ and $S_{DR}(G)_m$ be the inverse images of $S(G)_a$ and $S(G)_m$ via the "edge morphisms" $H^1_{DR}(A) \to H^1(\mathcal{O}_A)$ and $H^1_{DR}(A)_m \to \mathrm{Pic}^0(A)$.

THEOREM 2. Let G be an irreducible commutative algebraic $\mathcal{U}$-group. Then G admits a structure of algebraic D - group if and only if

$$\ell\nabla(S_{DR}(G)_m) \subset S_{DR}(G)_a \qquad \text{and}$$

$$\nabla(S_{DR}(G)_a) \subset S_{DR}(G)_a$$

In particular if G is the universal extension E(A) of an abelian $\mathcal{U}$- variety A by a vector group then G has a structure of algebraic D-group (recall that E(A) is an extension of A by G_a^g, $g = \dim A$, having no affine quotient); this consequence of Theorem 2 is also a consequence of the Grothendieck-Messing-Mazur crystalline theory [MM]. Note that the algebraic D - group $G(\Gamma)$ associated to the Δ_o - group derived from the equation (3b) above is a special case of this construction! Note also that Theorem 2 says more generally that any extension of E(A) by a torus G_m^N has a structure of algebraic D - group !

Theorem 2 will be deduced from a general duality theorem relating the Gauss-Manin connection and its multiplicative analogue to the "adjoint connection" on the Lie algebra of commutative algebraic D - groups. Our duality theorem generalizes certain aspects of the theory in [MM] and [BBM] and our proof is quite "elementary" (although computational !).

Let's pass to discussing "problem 2)" of "describing $P(G/\mathcal{U})$.

It is "well known" that the automorphism functor

$$\underline{\mathrm{Aut}}\, G : \{\mathcal{U}\text{- schemes}\} \to \{\text{groups}\}$$

$$(\underline{\mathrm{Aut}}\, G)(S) = \mathrm{Aut}_{S\text{-grsch.}}(G \times S)$$

of an algebraic $\mathcal{U}$- group G is not representable in general [BS] (by the way we will prove in Chapter 4 that the restriction of $\underline{\mathrm{Aut}}\, G$ to {reduced $\mathcal{U}$- schemes} is representable by a locally algebraic $\mathcal{U}$- group Aut G this providing a positive answer to a question of Borel and Serre [BS] p. 152). Then $P(G/\mathcal{U})$ is obviously identified with the Lie algebra $L(\underline{\mathrm{Aut}}\, G)$ of the functor $\underline{\mathrm{Aut}}\, G$, which contains the Lie algebra L(Aut G), but in general exceeds it (due to "nonrepresentability of $\underline{\mathrm{Aut}}$ over non-reduced schemes"). If G is commutative $P(G/\mathcal{U})$ is easily identified with $X_a(G) \otimes L(G)$ (where $\mathrm{Der}_{\mathcal{U}} \mathcal{O}_G$ is identified with $\mathcal{O}(G) \otimes L(G)$). For noncommutative G the analysis of $P(G/\mathcal{U})$ will be quite technical; we will perform it in some detail for G affine and get complete results in some special cases (e.g. in case the radical of G is nilpotent or if the unipotent radical of G is commutative), cf. Chapter 2.

Let's discuss in what follows the "splitting" problem for algebraic D-groups. If G_o is any irreducible algebraic $\mathcal{K}$- group one can construct an algebraic D - group $G = G_o \otimes_{\mathcal{K}} \mathcal{U}$ by letting D act on $\mathcal{O}_G$ via $1 \otimes \delta$; any algebraic D - group isomorphic to one obtained in this way will be called split. A Δ_o - group will be called split if it is isomorphic to $\check{G}_o$ for some algebraic

$\mathcal{K}$- group G_o; this is equivalent to saying that $G(\Gamma)$ is split. For instance the Δ_o - group Γ defined by equation (1) above is split; for if $\gamma_1, \dots, \gamma_n \in \mathcal{U}$ is a fundamental system of solutions of (1) then we have an isomorphism of Δ - groups

$$(G^n_{a,\mathcal{K}})^\vee = \mathcal{K}^n \to \Gamma$$

$$(c_1, \dots, c_n) \to \Sigma c_i \gamma_i$$

On the contrary the Δ_o - groups derived from equations (2) and (3) are not split; for (3b) this is clear because $G(\Gamma)$ does not descend to $\mathcal{K}$ while for (2) this follows because the D-submodule of $\mathcal{U}[\chi,\chi^{-1},x]$ generated by χ is infinite dimensional. Now for any algebraic $\mathcal{U}$- group the set of split algebraic D - group structures on G can be proved to be a principal homogenous space for L(Aut G) so, at least in case G descends to $\mathcal{K}$, the problem of determining what algebraic D - group structures on G are split is equivalent to determining what elements of $L(\underline{\mathrm{Aut}}\, G)$ actually lie in L(Aut G). In particular if G descends to $\mathcal{K}$ and $\underline{\mathrm{Aut}}$ G is representable then any algebraic D - group structure on G is split; this is the case if G is linear reductive or unipotent. More generally we will prove in Chapter 4 the following:

THEOREM 3. For an algebraic D-group G the following are equivalent:

1) G is split.

2) δ preserves the (ideal sheaf of the) unipotent radical U of the linear part of G.

3) δ preserves the (ideal sheaf of) $U/U \cap [G,G]$ in $G/[G,G]$.

4) δ preserves the maximum semiabelian subfield of the function field $\mathcal{U}(G)$ (which by definition is the function field of the maximum semiabelian quotient of G; recall that "semiabelian" means "extension of an abelian variety by a torus").

Consequently a Δ_o - group Γ is split iff $\Gamma/[\Gamma,\Gamma]$ is so, iff δ preserves the maximum semiabelian subfield of $\mathcal{U}<\Gamma>$; if in addition Γ is a Δ - closed subgroup of $\widehat{GL}_N$ for some $N \geq 1$ and we put $I := \{$ideal of all functions in $\mathcal{U}\{y_{ij}\}_d$, $d = \det(y_{ij})$, vanising on $\Gamma\}$ and $\mathcal{U}\{\Gamma\} := \mathcal{U}\{y_{ij}\}_d/I$ then Γ is split iff all group-like elements of the Hopf algebra $\mathcal{U}\{\Gamma\}$ are killed by δ.

We would like to close this presentation of our main results with a theorem about Δ - subgroups of abelian varieties; we need one more definition. We say that a Δ_o - group Γ has no non-trivial linear representation if any Δ - regular homomorphism $\Gamma \to \widehat{GL}_N$ is trivial.

THEOREM 4. Let A be an abelian $\mathcal{U}$-variety of dimension g. Then there is a unique Δ_o - subgroup $A^{\#}$ with the following properties:

1) it is Zariski dense in $\hat{A}$,

2) it has no non-trivial linear representation.

Moreover as A varies in the modului space $\mathcal{A}_{g,n}$ of principally polarized abelian $\mathcal{U}$- varieties with level n - structure, $n \geq 3$, the function $A \mapsto \dim A^{\#}$ varies lower semicontinuously with respect to the Δ - topology of $\hat{\mathcal{A}}_{g,n}$ and assumes all values between g and 2g.

In case g = 1 one may say more namely:

a) either A descends to $\mathcal{K}$, say $A = A_o \otimes_{\mathcal{K}} \mathcal{U}$, and then $A^{\#} = A_o(\mathcal{K})$ so $\dim A^{\#} = 1$.

b) or A does not descend to $\mathcal{K}$ and in this case $\dim A^{\#} = 2$.

Note that the Δ - group Γ defined by (3b) is nothing but the group $A^{\#}$ above. On the other hand the Δ - group Γ defined by (3a) is an extension of $(G_{a,\mathcal{K}})^{\vee} \simeq \mathcal{K}$ by $A^{\#}$ above.

6. Amplifications

Most definitions given so far in our Introduction will be "enlarged" in the body of the text and most results will be proved in a generalized form. For instance the definitions related to Δ - manifolds and Δ - groups will be given for "partial" rather than "ordinary" differential fields, i.e. for fields $\mathcal{U}$ equiped with several commuting derivations. Algebraic D - groups will be allowed to be reducible and will be defined for D = K[P] any "k-algebra of linear differential operators on a field extension K of k" which is "built on a Lie K/k-algebra P" (cf. [NW] or Section 0 below); this degree of generality and abstractness might seem excessive, but it is motivated in many ways:

1) our wish to deal with Δ - groups over partial differential universal fields

2) our wish to "compute", for a given algebraic K-group G (K algebraically closed of characteristic zero, containing a field k) the smallest algebraically closed field of definition K_G of G between k and K; the existence of K_G was proved in [B_5] but we will reprove this in a different way in Chapter 4.

3) our wish to relate our topic to Deligne's "regular D-modules".

4) our wish to take a (shy) look at "algebraic D-groups in positive characteristic".

The reader will appreciate the usefulness of this more general concept of algebraic D-group in the light of the arguments 1)-4) above.

7. Plan

The book opens with a preliminary section 0 which fixes terminology and conventions. In Chapter 1 we present the main concepts related to algebraic D-groups. Chapters 2 and 3 are devoted to affine, respectively to commutative algebraic D-groups. Chapter 4 deals with algebraic D-groups which are not necessarily affine or commutative. Chapter 5 opens with an introduction to Δ - manifolds and Δ - groups and then deals with the "structure" and "moduli" of Δ_o - groups.

Internal references to facts not contained in section 0 will be given in the form (X,y,z) or simply (X,y) where X is the number of the chapter and y is the number of the section. Within the same chapter X we write (y,z) instead of (X,y,z). References to section 0 will be given in the form (0,z). Each chapter will begin with a brief account of its contents and of its specific conventions.

8. Prerequisites

The reader is assumed to have only some basic knowledge of algebraic geometry [Har] and of algebraic groups [H], [Ro], [Se]. The non-experts in these fields might still appreciate the results of this book via our comments and various examples.

No knowledge of the Kolchin-Cassidy theory [K_i][C_j] is assumed; the elements of this theory which are relevant for our approach will be quickly reviewed in the book.

Finally one should note that the present book is ideologically a continuation of our previous book [B_1]; but it is logically independent of it.

9. Aknowledgements

A few words cannot express the thanks I owe to Ellis Kolchin and the members of his Columbia Seminar: P. Cassidy, R. Cohn, L. Goldman, J. Johnson, S. Morison, P. Landesman, W. Sit. Their continuous encouragment, constant support and numerous suggestions played an important role in my fulfilling the task of writing this book.

I would also like to aknowledge my debt to H. Hamm and O. Laudal for useful conversations.

Last but not least I would like to thank Camelia Minculescu for her excellent typing job.

June 1989

Revised: June 1991

A. Buium

CONTENTS

0. TERMINOLOGY AND CONVENTIONS

(0.1) Unless otherwise stated rings, fields and algebras are generally assumed associative commutative with unit. This will not apply however to Lie algebras, universal enveloping algebras or to Hopf algebras (the latter are understood in the sense of Sweedler [Sw]).

(0.2) Terminology of algebraic geometry is the standard one (cf. for instance [Har][DG]). Nevertheless we make the convention that all schemes appearing are separated. For any morphism of schemes $f : X \to Y$ we generally denote the defining sheaf morphism $\mathcal{O}_Y \to f_* \mathcal{O}_X$ by the same letter f. If both X and Y are schemes over some field K then by a K- f-derivation we mean a K-derivation of $\mathcal{O}_Y$ into the $\mathcal{O}_Y$-module $f_* \mathcal{O}_X$. By a K-variety (K a field) we will always mean a (separated!) geometrically integral K-scheme of finite type. By a (locally) algebraic K-group we will mean a geometrically reduced (locally) algebraic K-group scheme. For any integral K-algebra A (respectively for any integral K-scheme X) we let K(A) (respectively K(X)) denote the quotient field of A (respectively of X). For any K-variety X and any field extension K_1/K we denote by $X(K_1)$ the set $\mathrm{Hom}_{K\text{-sch}}(\mathrm{Spec}\, K_1, X)$ of K_1-points of X; more generally we write X(Y) instead of $\mathrm{Hom}_{K\text{-sch}}(Y,X)$ for any K-schemes X,Y. A morphism $X \to Y$ of K-varieties will be called surjective if the map $X(K) \to Y(K)$ is surjective. For any K-scheme X, T_X will denote the sheaf $\mathrm{Der}_K \mathcal{O}_X$. If $X = \mathrm{Spec}\, A$ for some K-algebra A we sometimes write Der(A/K) instead of $\mathrm{Der}_K A$.

(0.3) All affine K-group schemes G are tacitly assumed to be such that the ring $\mathcal{O}(G)$ of global regular functions is at most countably generated as a K-algebra. This is a harmless assumption in view of our applications in Chapter 5.

The unit in any K-group scheme G will be denoted by e; if G is commutative we will sometimes write 0 instead of e.

For any locally algebraic K-group G, G^o will denote the identity component; Z(G) will denote the center and we put $Z^o(G) := (Z(G))^o$.

(0.4) For any algebraic $\mathbb{C}$ - variety X (respectively algebraic $\mathbb{C}$ - group G) we denote by X^{an} (resp. G^{an}) the associated analytic space (respectively the analytic Lie group). For any analytic manifold $\mathfrak{X}$ we denote by $T_{\mathfrak{X}}$ the analytic tangent bundle.

(0.5) Our terminology of differential algebra is a combination of terminologies from $[C_i]$ $[K_i]$ [NW] and $[B_i]$. In what follows we shall review it in some detail and also introduce some new concepts.

Let K/k be a field extension. By a Lie K/k - algebra [NW] we mean a K-vector space P which is also a Lie k-algebra, equipped with a K-linear map $\partial : P \to \mathrm{Der}_k K$ of Lie k-algebras such that

$$[p_1,\lambda p_2] = \partial(p_1)(\lambda)p_2 + \lambda[p_1,p_2] \quad \text{for} \quad \lambda \in K,\ p_1,p_2 \in P$$

Let's give some basic examples of Lie K/k - algebras.

EXAMPLE 1. Start with a derivation $d \in Der_k K$; one can associate to it a Lie K/k - algebra P of dimension 1 over K by letting

$$P = Kp$$

$$\partial(\lambda p) = \lambda d, \quad \lambda \in K$$

$$[\lambda_1 p, \lambda_2 p] = (\lambda_1 d\lambda_2 - \lambda_2 d\lambda_1)p, \qquad \lambda_1, \lambda_2 \in K$$

So we call the attention on the fact that this P is far from being commutative!

EXAMPLE 2. Start now with a family of derivations $(d_i)_{i\in I}$, $d_i \in Der_k K$. Then one can associate to it the "free Lie K/k-algebra" P built on this family: by definition P has the property that it contains a family of elements $(p_i)_{i\in I}$ with $\partial(p_i) = d_i$ such that for any K/k - algebra P' and any family $(p_i')_{i\in I}$, $p_i' \in P'$ with $\partial'(p_i') = d_i$ there is a unique Lie K/k - algebra map $f : P \to P'$ with $f(p_i) = p_i'$. We leave to the reader the task of constructing this P. Note that if I consists of one element this P is the same with the one in Example 1.

EXAMPLE 3. Assume we are given a family $(d_i)_{i\in I}$, $d_i \in Der_k K$ such that $[d_i, d_j] = 0$ for all $i,j \in I$. Then one can construct a new Lie K/k-algebra P as follows: we let P have a K-basis $(p_i)_{i\in I}$ we let $\partial(p_i) = d_i$ and define the bracket [,] by the formula

$$[\lambda p_i, \mu p_j] = \lambda(d_i\mu)p_j - \mu(d_j\lambda)p_i, \qquad \lambda, \mu \in K, \ i,j \in I$$

This P can be called the "free integrable Lie K/k-algebra" built on our family of derivations. Once again if I consists of one element, this P coincides with the one in Example 1.

EXAMPLE 4. Let $k \subset E \subset K$ be an intermediate field. Then $P = Der_E K$ together with its inclusion $\partial : Der_E K \to Der_k K$ is an example of Lie K/k - algebra.

Other remarkable examples of Lie K/k - algebras will appear in (I.1).

Given such a P one associates to it [NW] a k-algebra of differential operators D = K[P]: by definition K[P] is the associative k-algebra generated by K and P subject to the relations

$\lambda p = v(\lambda, p)$ for $\lambda \in K$, $p \in P$ where $v : K \times P \to P$ is the vector space structure map

$p\lambda - \lambda p = \partial(p)(\lambda)1$ for $\lambda \in K$, $p \in P$

$p_1 p_2 - p_2 p_1 = [p_1, p_2]$ for $p_1, p_2 \in P$

Recall [NW] that D = K[P] has a K-basis consisting of all the monomials of the form $p_{i_1}^{e_1} \dots p_{i_n}^{e_n}$ where $e_j \geq 0$ and $i_1 \leq \dots \leq i_n$ in some total order on a basis $(p_i)_i$ of P. For instance if P is 1-dimensional, P = Kp, as in Example 1 above then

$$D = \sum_{i\geq 0} Kp^i \qquad \text{(direct sum)}$$

is the ring of "linear ordinary differential operators" on K "generated by K and p"; multiplication in D corresponds to "composition of linear differential operators". This example is quite familiar: indeed, assume $k = \mathbb{C}$, $K = \mathbb{C}(t)$ is the field of rational functions and $\partial(p) = d/dt$; then $D = \mathbb{C}(t)[d/dt]$ is nothing but the "rational" Weyl algebra. Similarity one can obtain the "rational" Weyl algebra in several variables $D = \mathbb{C}(t_1,\dots,t_n)[d/dt_1,\dots,d/dt_n]$ starting with P as in Example 3 above. Since D is a ring we may speak about D-modules (always assumed to be left modules). If V is a D-module we define its set of P-constants $V^D = \{x \in V;\ px = 0 \text{ for all } p \in P\}$; it is a vector space over the field $K^D = \{\lambda \in K;\ px = 0 \text{ for all } p \in P\}$. We sometimes write also V^P and K^P instead of V^D and K^D and speak about P-modules instead of D-modules. Recall from [NW] that if V,W are D- modules then $V \oplus W$, $V \otimes_K W$ and $\mathrm{Hom}_K(V,W)$, hence in particular $V^0 = \mathrm{Hom}_K(V,K)$, have natural structures of D - modules and the following formulae hold

$$p(x,y) = (px,py) \quad \text{for } p \in P,\ (x,y) \in V \oplus W$$

$$p(x \otimes y) = px \otimes y + x \otimes py \quad \text{for } p \in P,\ x \otimes y \in V \otimes_K W$$

$$(pf)(x) = p(f(x)) - f(px) \ \text{ for } p \in P,\ x \in V,\ f \in \mathrm{Hom}_K(V,W)$$

By a D - algebra (respectively associative D - algebra, Lie D - algebra, Hopf D - algebra) we understand a D - module A which is also a K - algebra (respectively an associative K - algebra, Lie K - algebra, Hopf K - algebra) in such a way that all structure maps are D - module maps; here by "structure maps" we understand multiplication $A \otimes_K A \to A$, unit $K \to A$, comultiplication $A \to A \otimes_K A$ and co-unit $A \to K$($A \otimes_K A$ and K are viewed with their natural D - module structure). A D - algebra is called D - finitely generated if it is generated as a K - algebra by some finitely generated D - submodule of it.

Note that if P is as in Example 1 (respectively 2, 3) a D - algebra is simply a K - algebra A together with a lifting of the derivation $d \in \mathrm{Der}_k K$ to a derivation $\tilde{d} \in \mathrm{Der}_k A$ (respectively with liftings of d_i to $\tilde{d}_i \in \mathrm{Der}_k A$ in case of Example 2 and with pairwise commuting liftings $\tilde{d}_i \in \mathrm{Der}_k A$ in case of Example 3).

A D - algebra which is a field will be called a D - field. Clearly K is a D - field. K will be called D - algebraically closed if for any D - finitely generated D - algebra A there exists a D - algebra map $A \to K$.

Of course one would like to "see" an example of D - algebraically closed field. Unfortunately we can't "show" any (although one can prove the existence of enough of them, see (0.12) below). But one should not forget that with a few exceptions a similar situation occurs with algebraically closed fields.

Note that if A_1, A_2 are D - algebras then $A_1 \otimes_K A_2$ becomes a D - algebra.

Clearly if A is a D - algebra then $\mathrm{Im}(P \to \mathrm{End}_k A) \subset \mathrm{Der}_k A$. If no confusion arrises we denote the image of any $p \in P$ in $\mathrm{Der}_k A$ by the same letter p.

(0.6) Following [B_1] we can define D - schemes: these are K - schemes X whose structure sheaf $\mathcal{O}_X$ is given a structure of sheaf of D - algebras (i.e. $\mathcal{O}_X(U)$ is a D - algebra for all open sets U and restriction maps $\mathcal{O}_X(U_1) \to \mathcal{O}_X(U_2)$ for $U_2 \subset U_1$ are D - algebra maps). For any

D - algebra A, Spec A is a D - scheme. Clearly D - schemes form a category (whose morphisms will be called D - morphisms or D - maps: these are the morphisms $f: X \to Y$ such that $\mathcal{O}_Y \to f_* \mathcal{O}_X$ is a morphism of sheaves of D - algebras). This category is easily seen to have fibre products. Indeed if X and Y are D - schemes over a D - scheme Z then the scheme fibre product $X \times_Z Y$ has a natural D - scheme structure coming from the D - module structure of tensor products (0.5); if $p \in P$ then the corresponding derivation on the structure sheaf of $X \times_Z Y$ will be denoted by $p \otimes 1 + 1 \otimes p$ (this convention agrees with formula "$p(x \otimes y) = px \otimes y + x \otimes py$" on products of affine pieces: $X_0 \times_{Z_0} Y_0$, X_0, Y_0, Z_0 affine open subsets of X, Y, Z). A D - scheme will be called of D - finite type if it can be covered with finitely many affine open sets U_i whose coordinate rings are D - finitely generated.

By a D - variety we understand a D - scheme X which is a K - variety; we shall usually write $K\langle X \rangle$ instead of K(X). Clearly $K\langle X \rangle$ is a D - field.

(0.7) Let A be a D - algebra; by a D - ideal we mean an ideal J which is a D - submodule of A. For any D - ideal the quotient A/J is a D - algebra. Let X be a D - scheme; by a closed D - subscheme we mean a subscheme Y whose sheaf of ideals is a sheaf of D-submodules of $\mathcal{O}_X$ (so Y is in particular a D-scheme). By a D-point on a D-scheme X we mean a (non necessarily closed) point $x \in X$ such that the maximal ideal of $\mathcal{O}_{X,x}$ is a D - ideal; we denote by X_D the set of all D - points of X. If X,Y are D - schemes we write $X_D(Y)$ instead of $\mathrm{Hom}_{D\text{-sch}}(Y,X)$; if $Y = \mathrm{Spec}\, A$ for some D - algebra A we write $X_D(A)$ instead of $X_D(Y)$. Elements of $X_D(A)$ are called D - A - points.

(0.8) Now comes our main concept. By a D - group scheme we shall mean a group object in the category of D - schemes i.e. a D - scheme G which is also a K - group scheme such that the multiplication $G \times_K G \to G$, the antipode $G \to G$ and the unit $\mathrm{Spec}\, K \to G$ are D - morphisms. By an algebraic D - group we will understand a D - group scheme which is also an algebraic K - gorup. (This is not the concept of algebraic D - group from [NW]!). One defines in an obvious way the notions of D - subgroup scheme and algebraic D - subgroup.

(0.9) If X and Y are two D - schemes then to give a D - map $f: X \to Y$ is equivalent to giving for any D-algebra A a map $f_A : X_D(A) \to Y_D(A)$ behaving functorially in A. If both X and Y are reduced it is sufficient to define f_A only for reduced A. In particular any D - group scheme G can be recaptured from the functor

$$\{D\text{ - algebras}\} \to \{\text{groups}\}, \qquad A \mapsto G_D(A)$$

(0.10) If K_1/K is a D - field extension (i.e. a morphism $K \to K_1$ of D - fields) then one easily checks that $P_1 := K_1 \otimes_K P$ has a natural structure of Lie K_1/k-algebra and that $D_1 := K_1[P_1]$ identifies with $K_1 \otimes_K D$. Moreover, if V is a D-module then $K_1 \otimes_K V$ has a natural structure of D_1-module. A similar assertion holds for algebras, associative algebras, Lie algebras, Hopf algebras, schemes, group schems, varieties, algebraic groups (instead of modules). If X is a D-variety we will often write $K_1\langle X \rangle$ instead of $K_1 \langle K_1 \otimes_K X \rangle$.

(0.11) An important remark is in order. Namely the above D-schemes and D-varieties are objects quite close to the Δ - schemes and Δ - varieties considered in $[B_1]$: more precisely Δ - schemes over a Δ - field K in the sense of $[B_1]$ coincide with D - schemes in the above sense where D=K[P], P = free Lie K/k - algebra on Δ, $k = K^{\Delta}$. But our algebraic D-groups above are objects quite different from (although, as we shall see, deeply related to) the Cassidy-Kolchin Δ - groups $[C_i]$$[K_i]$. The relation between the latter concepts will be discussed in Chapter 5. Finally the algebraic D-groups from [NW] are precisely our affine D-group schemes!

One should also note that Kolchin's Δ - fields $[K_1]$$[K_2]$ correspond in our language to fields K equiped with a Lie K/k-algebra P which has a finite commuting K-basis.

(0.12) Kolchin proved in $[K_3]$ that if P is a Lie K/k-algebra having a finite commuting K-basis (char K = 0) and D = K[P] then there exists a D-field extension K_1/K such that K_1 is D_1-algebraically closed, $D_1 = K_1 \otimes_K D$.

For the reader familiar with $[K_3]$ note that K is D-algebraically closed iff (viewed as a Δ - field) it is constrainedly closed in Kolchin's sense.

(0.13) Various other standard definitions can be given in the category of D-schemes. We shall need for instance the concept of left (respectively right) D-action of a D-group scheme G on a D-scheme X: it is of course a D-map $G \times X \to X$ (respectively $X \times G \to X$) satisfying the usual axioms of an action. Similarily, one can define D-principal homogeneous spaces (these are right D-actions $X \times G \to X$ such that the induced D-map $X \times G \to X \times X$ is an isomorphism). One can speak moreover about D-actions of a D-group scheme G on another D-group scheme H and about D-cocycles of G in H. It is useful to have some notations also. We let $PHS_D(G/K)$ be the set of D-isomorphism classes of D-principal homogeneous spaces for a D-group scheme G; it is a pointed set with distinguished element called 1 and represented by G itself. Any D-principal homogeneous space V representing 1 will be called trivial; V is trivial if and only if $V_D(K) \neq \emptyset$.

(0.14) Let's "recall" from $[B_1]$ Chapter 1 some useful elementary facts about D-schemes. Let X be a D-scheme, X noetherian.

1) If X is reduced, its irreducible components are D-subschemes.

2) If Y is an integral subscheme of X, its generic point is in X_D if and only if Y is a D-subscheme.

3) If char K = 0 then X_{red} is a D-subscheme of X. In particular by 1), 2) above $X_D \neq \emptyset$. Assume moreover that K is D-algebraically closed and X is of D-finite type. Then $X_D(K) \neq \emptyset$.

4) If char K = 0 and X is a D-variety then for any maximal element x in X_D the extension $K(x)^D/K^D$ is algebraic (where K(x) = residue field of X at x).

5) If $f: X \to Y$ is a faithfully flat morphism of integral K-schemes and K(Y) is a D-subfield of $K\langle X \rangle$ then Y has a (unique!) structure of D-scheme such that f is a D-map.

6) If $f: X' \to X$ is an étale morphism of K-varieties then X' has a unique structure of D-variety such that f is a D-map.

7) If $f : X \to Y$ is a D-map of D-schemes and $y \in Y_D$ then the fibre $f^{-1}(y)$ has a natural structure of D-scheme.

8) Assume V is a normal K-variety such that K(V) has a structure of D-field, D = K[P]. Assume moreover that $\dim_K P < \infty$. Then the locus $V_o = \{x \in V;\ D\mathcal{O}_{V,x} \subset \mathcal{O}_{V,x}\}$ is Zariski open (so it is a D-scheme) and $V \setminus V_o$ is the support of some divisor on V.

Recall also that we have proved in $[B_1]$ p. 28 the following:

(0.15) THEOREM. Assume K is algebraically closed, and X is a projective D-variety. Then K^D is a field of definition for X (i.e. X is K-isomorphic to $X_o \otimes_{K^D} K$ for some projective K^D-variety X_o).

(0.16) We close by introducing a concept which will appear several times in our work. Let P be Lie K/k algebra, S_m a group and S_a a Lie K-algebra on which S_m acts by Lie K-algebra automorphisms. By a logarithmic P-connection on (S_m, S_a) we mean a pair $(\nabla, \ell\nabla)$ of K-linear maps

$$\nabla : P \to \mathrm{Der}_k(S_a, S_a) \qquad p \mapsto \nabla_p$$

$$\ell\nabla : P \to Z^1(S_m, S_a) \qquad p \mapsto \ell\nabla_p$$

such that $\nabla_p(\lambda x) = \lambda \nabla_p x + \partial(p)(\lambda) x$ for all $p \in P$, $\lambda \in K$, $x \in S_a$. Now $Z^1(S_m, S_a)$ has a natural structure of Lie K-algebra induced from that of S_a. We say that $(\nabla, \ell\nabla)$ is integrable if for any $p_1, p_2 \in P$ the following formulae hold:

(0.16.1) $$\nabla_{[p_1,p_2]} = [\nabla_{p_1}, \nabla_{p_2}]$$

(0.16.2) $$\ell\nabla_{[p_1,p_2]} = \nabla_{p_1} \circ \ell\nabla_{p_2} - \nabla_{p_2} \circ \ell\nabla_{p_1} + [\ell\nabla_{p_1}, \ell\nabla_{p_2}]$$

Of course condition 1) is equivalent to saying that there exists a (unique) D-module structure on S_a (D = K[P]) such that $px = \nabla_p x$ for all $p \in P$ and $x \in S_a$.

A "trivial" example of integrable logarithmic $\mathrm{Der}_k K$-connection is on (K^*, K) where K^* acts trivially on K and K is viewed as an abelian K-algebra: it is given by the formulae $\nabla_p x = px$, $\ell\nabla_p y = y^{-1} py$ for all $p \in \mathrm{Der}_k K$, $x \in K$, $y \in K^*$. Various significant examples of logarithmic connections will appear in (I.1) and (III.1).

Note that if in our definitions S_m is trivial and S_a is an abelian Lie algebra then a logarithmic P-connection is simply a P-connection on S i.e. a K-linear map $\nabla : P \to \mathrm{Hom}_k(S,S)$ satisfying the usual Leibnitz rule.

CHAPTER 1. FIRST PROPERTIES

Everywhere in this chapter K/k is a field extension with K algebraically closed (of arbitrary characteristic), P is a Lie K/k-algebra and D = K[P] (cf. Section 0). In section 1 of this chapter we associate to any K-group scheme G a basic Lie K/k-algebra P(G) and investigate its "first properties". Sections 2 and 3 deal with basic concepts of prolongations, local finiteness and splitting.

1. Basic spaces and maps

(1.1) Let G be a K-group scheme. We shall denote by $\mu : G \times G \to G$, $S : G \to G$, $\varepsilon : \operatorname{Spec} K \to G$ the multiplication, antipode and unit; we will also denote by

$$\mu : \mathcal{O}_G \to \mu_* \mathcal{O}_{G\times G}$$

$$S : \mathcal{O}_G \to S_* \mathcal{O}_G$$

$$\varepsilon : \mathcal{O}_{G,e} \to K$$

the induced comultiplication, antipode and co-unit (cf. (0.2)). Define

$$\mathrm{Der}_k G = \{\text{space of all k-derivations of } \mathcal{O}_G \text{ into itself}\}$$

$$L(G) = \{v \in \mathrm{Der}_k G;\ v \text{ is K-linear and } \mu v = (v \otimes 1)\mu\}$$

$$P(G) = \{p \in \mathrm{Der}_k G;\ pK \subset K,\ \mu p = (p \otimes 1 + 1 \otimes p)\mu,\ pS = Sp,\ \varepsilon p = p\varepsilon\}$$

Clearly $\mathrm{Der}_k G$ is a K-linear space and a Lie k-algebra. Moreover L(G) is a Lie K-algebra and P(G) is a Lie K/k-algebra. Both L(G) and P(G) are Lie k-subalgebras of $\mathrm{Der}_k G$. Define D(G) = K[P(G)]. Moreover for any $p \in P(G)$ let K[p] be the k-algebra of differential operators associated to the 1-dimensional Lie K/k-algebra Kp (clearly $K[p] \subset K[P]$).

Note that we made a slight abuse in our definitions. For instance the formula $\mu p = (p \otimes 1 + 1 \otimes p)\mu$ in the definition of P(G) means that the following diagram of abelian sheaves on G is commutative:

$$\begin{array}{ccc} \mathcal{O}_G & \xrightarrow{\ \mu\ } & \mu_* \mathcal{O}_{G\times G} \\ {\scriptstyle p}\downarrow & & \downarrow{\scriptstyle \mu_*(p \otimes 1 + 1 \otimes p)} \\ \mathcal{O}_G & \xrightarrow[\ \mu\ ]{} & \mu_* \mathcal{O}_{G\times G} \end{array}$$

where $p \otimes 1 + 1 \otimes p$ was defined in (0.6). On the other hand the formula $\mu v = (v \otimes 1)\mu$ in the definition of L(G) means that the diagram

$$\begin{array}{ccc} \mathcal{O}_G & \xrightarrow{\mu} & \mu_*\mathcal{O}_{G\times G} \\ p\downarrow & & \downarrow \mu_*(v\otimes 1) \\ \mathcal{O}_G & \xrightarrow[\mu]{} & \mu_*\mathcal{O}_{G\times G} \end{array}$$

is commutative where $v\otimes 1$ is the K-derivation of $\mathcal{O}_{G\times G}$ acting as v on the "first component" and vanishing on the "second component". Such abuses greatly simplify our computations and will be done from now on systematically.

Note that if char $K > 0$ then $\mathrm{Der}_k G = \mathrm{Der}_K G$ because K is perfect. We can define more generally for any intermediate field K_o between k and K the spaces $\mathrm{Der}_{K_o} G$ and $P(G/K_o)$ as consisting of all K_o-linear members of $\mathrm{Der}_k G$ and P(G); clearly $P(G/K_o)$ is a Lie K/K_o-algebra and $P(G/K) = \mathrm{Ker}(\partial : P(G) \to \mathrm{Der}_k K)$.

Apriori it is not clear that L(G) can be identified with the Lie algebra Lie G of G as defined in [DG] (Lie G = tangent space of G at e). But of course, if G is an algebraic group, L(G) coincides with the Lie algebra of right invariant derivations in $\mathrm{Der}_K G$ and hence with Lie G. We will check below that L(G)=Lie G at least in one more case namely when G is integral, affine (see (1.8)).

As for one interpretation of P(G) note that there is a natural bijection between the set of D-group scheme structures on G and the set

$$\mathrm{Hom}_{\mathrm{Lie}\,K/k\text{-alg}}(P,P(G))$$

In particular G has a structure of D(G)-group scheme and this structure is "universal" in an obvious sense. This remark is basic for our work. We will usually investigate first this "universal" D(G)-structure and then "specialize" to arbitrary D-structures.

EXAMPLE. Assume $P = Kp$ is 1-dimensional; then the set of D-group scheme structures on G is in bijection with the set $P(G/\partial(p))$ of all derivations $\tilde{p}$ in P(G) inducing the derivation $\partial(p)$ on K. So $P(G/\partial(p))$ is a principal homogenous space for P(G/K).

Before computing P(G) for a few specific G's we need to prove a series of general properties.

(1.2) LEMMA. $[P(G), L(G)] \subset L(G)$ where the bracket is taken in $\mathrm{Der}_k G$.

Proof. Let $p \in P(G)$, $v \in L(G)$. We have:

$$\mu p v = (p\otimes 1 + 1\otimes p)\mu v = (p\otimes 1 + 1\otimes p)(v\otimes 1)\mu = (pv\otimes 1 + v\otimes p)\mu$$

$$\mu v p = (v\otimes 1)\mu p = (v\otimes 1)(p\otimes 1 + 1\otimes p)\mu = (vp\otimes 1 + v\otimes p)\mu$$

hence

$$\mu[p,v] = ([p,v]\otimes 1)\mu$$

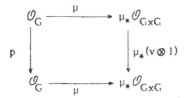

which proves the lemma.

We denote in what follows by G_m, G_a the multiplicative and additive 1-dimensional algebraic K-groups. Recall that

$$G_m = \mathrm{Spec}\, K[\chi,\chi^{-1}]$$

$$\mu\chi = \chi\otimes\chi \in K[\chi,\chi^{-1}]\otimes K[\chi,\chi^{-1}]$$

$$S\chi = \chi^{-1}$$

$$\varepsilon\chi = 1$$

and that

$$G_a = \mathrm{Spec}\, K[\xi]$$

$$\mu\xi = \xi\otimes 1 + 1\otimes\xi \in K[\xi]\otimes K[\xi]$$

$$S\xi = -\xi$$

$$\varepsilon\xi = 0$$

For any K-group scheme we G put

$$X_m(G) = \mathrm{Hom}_{K\text{-grsch}}(G,G_m)$$

$$X_a(G) = \mathrm{Hom}_{K\text{-grsch}}(G,G_a)$$

Clearly $X_m(G)$ is a subgroup of $\mathcal{O}^*(G)$ and $X_a(G)$ is a K-linear subspace of $\mathcal{O}(G)$.

(1.3) LEMMA. For any $p \in P(G)$, $\chi \in X_m(G)$, $\xi \in X_a(G)$ and $v \in L(G)$ we have $\chi^{-1}p\chi \in X_a(G)$, $p\xi \in X_a(G)$ and $v\xi \in K$.

Proof. The following holds: $\mu p\chi = (p\otimes 1 + 1\otimes p)\mu\chi = (p\otimes 1 + 1\otimes p)(\chi\otimes\chi) = p\chi\otimes\chi + \chi\otimes p\chi$. Multiplying the above equality by $\chi^{-1}\otimes\chi^{-1}$ we get

$$\mu(\chi^{-1}p\chi) = \chi^{-1}p\chi\otimes 1 + 1\otimes\chi^{-1}p\chi$$

i.e. $\chi^{-1}p\chi \in X_a(G)$.

Similarily one checks that $p\xi \in X_a(G)$.

Finally we have

$$\mu v\xi = (v\otimes 1)\mu\xi = (v\otimes 1)(\xi\otimes 1 + 1\otimes\xi) = v\xi\otimes 1$$

Applying $\varepsilon\otimes 1$ to the above equality we get

$$v\xi = (\varepsilon\otimes 1)\mu v\xi = (\varepsilon\otimes 1)(v\xi\otimes 1) = \varepsilon(v\xi) \in K$$

and the lemma is proved.

(1.4) We have an integrable logarithmic P(G)-connection $(\nabla, \ell\nabla)$ on $(\mathcal{O}^*(G), \mathcal{O}(G))$ defined by the formulae $\nabla_p f = pf$ and $\ell\nabla_p g = g^{-1}pg$ for all $p \in P(G)$, $f \in \mathcal{O}(G)$, $g \in \mathcal{O}^*(G)$.

Here $\mathcal{O}(G)$ is viewed as an abelian Lie K-algebra on which $\mathcal{O}^*(G)$ acts trivially. By (1.3) above this logarithmic connection induces an integrable logarithmic P(G)-connection $(\nabla, \ell\nabla)$ on $(X_m(G), X_a(G))$ where $X_a(G)$ is viewed as an abelian Lie K-algebra on which $X_m(G)$ acts trivially.

In what follows we prepare ourselves to deal with L(G) for G affine and with P(G/K) for G commutative and either affine or algebraic over K.

(1.5.) Assumed G is an affine K-group scheme. Then we define a linear topology on $Der_k G$ as follows. For any system of elements $f_1, \dots f_n \in$ (G) define

$$V_{f_1,\dots,f_n} = \{p \in Der_k G;\ pf_1 = \dots = pf_n = 0\}$$

Then we take as a fundamental system of open neighbourhoods of the origin in $Der_k G$ the above spaces $V_{f_1,\dots,f_n}$ as $f_1, \dots, f_n$ and n vary. Clearly $Der_k G$ with this topology is separated and complete (due to our convention made in (0.3) this topology has a countable basis).

(1.6) Recall from [DG] that (with our convention in (0.3)) any affine K-group scheme G is the projective limit $G = \varprojlim G_i$ of an inverse system

$$\dots \to G_{i+1} \xrightarrow{u_i} G_i \to \dots \to G_1$$

such that each u_i is a faithfully flat morphism of affine algebraic K-groups. In particular

$$\mathrm{Lie}\, G = \varprojlim \mathrm{Lie}\, G_i = \varprojlim L(G_i)$$

(1.7) Let

$$\dots \to V_{i+1} \to V_i \to \dots \to V_1$$

be a projective system of finite dimensional K-vector spaces with surjective connecting maps and let $V = \varprojlim V_i$ be viewed with its projective limit topology. Then one can find a sequence $v_1, v_2, \dots \in V$ (we agree to assume that the sequence is finite if V has finite dimension) such that for each index i there exists an integer n_i having the property that upon letting $\bar{v}_j$ be the image of v_j in V_i, $\bar{v}_1, \dots, \bar{v}_{n_i}$ is a K-basis in V_i and $\bar{v}_j = 0$ for all $j > n_i$. Call such a sequence $(v_i)_i$ a pseudo-basis for V. Clearly any element in V can be written uniquely as a convergent series $\Sigma\lambda_i v_i$ with $\lambda_i \in K$. If $\dim_k V < \infty$ a pseudo-basis is simply a basis.

(1.8) PROPOSITION. Let G be an affine integral K-group scheme and let $G = \varprojlim G_i$ as in (1.6). Then:

1) For any intermediate field K_0 between k and K the spaces $Der_{K_0} G$, $P(G/K_0)$ and L(G) are closed in $Der_k G$ hence they are complete.

2) Let $v_1, v_2, \dots$ be a pseudo-basis in $\varprojlim L(G_i)$ (we may view each v_i as an element of

$\mathrm{Der}_k G$). Then any element in $\mathrm{Der}_K G$ can be uniquely written as $\Sigma a_i v_i$ with $a_i \in$ (G) (which is a convergent series in $\mathrm{Der}_k G$).

3) An element $\Sigma a_i v_i$ as in 2) belongs to L(G) if and only if $a_i \in K$ for all i; in particular $L(G) = \varprojlim L(G_i)$, hence L(G) identifies with Lie G.

4) Assume G is commutative. Then an element $\Sigma a_i v_i$ as in 2) belongs to P(G/K) if and only if $a_i \in X_a(G)$ for all i.

Proof. 1) is easy and we omit proof.

2) Let $f_i : G \to G_i$ be the i-th canonical projection.

We have

$$\mathrm{Der}_K G = \varprojlim \mathrm{Der}_K(\mathcal{O}_{G_i}, f_{i_*}\mathcal{O}_G) = \varprojlim \mathrm{Hom}(\Omega_{G_i/K}, f_{i_*}\mathcal{O}_G) =$$

$$= \varprojlim H^0(G_i,(f_{i_*}\mathcal{O}_G) \times L(G_i)) = \varprojlim(\mathcal{O}(G) \times L(G_i))$$

which immediately implies assertion 2).

3) Clearly $\varprojlim L(G_i) \subset L(G)$ so $\Sigma\lambda_i v_i \in L(G)$ if $\lambda_i \in K$. Conversely, assume $v = \Sigma a_i v_i$, $\mu v = (v \otimes 1)\mu$. Then we have

$$\Sigma\mu(a_i)\mu v_i = \Sigma(a_i \otimes 1)(v_i \otimes 1)\mu = \Sigma(a_i \otimes 1)\mu v_i$$

We shall be done if we check that an equality of the type $\Sigma m_i \mu v_i = 0$ with $m_i \in \mathcal{O}(G \times G)$ implies $m_i = 0$ for all i. For each i we may find elements $f_1, \dots f_{n_i}$ in the maximal ideal m_i of $\mathcal{O}_{G_i,e}$ providing a K-basis in m_i/m_i^2 and such that $\varepsilon v_p f_q = \delta_{pq}$ (Kronecker delta) for all $1 \leq p,q \leq n_i$. Then $\det(v_p f_q)_{1\leq p,q\leq n_i} \in \mathcal{O}^*_{G_i,e}$. We get

$$\sum_{p=1}^{n_i} m_p \mu v_p f_q = 0 \quad \text{for} \quad 1 \leq q \leq n_i$$

Now

$$\det(\mu v_p f_q) = \mu(\det(v_p f_q)) \in \mathcal{O}^*_{G_i \times G_i, e\times e}$$

which implies that $m_p = 0$ for all $1 \leq p \leq n_i$.

4) If G is commutative so are all G_i's so for all i we have

$$\mu v_i = (v_i \otimes 1)\mu = (1 \otimes v_i)\mu \quad \text{and}$$

$$S v_i S = -v_i$$

Now for $p = \Sigma a_i v_i$ we have

$$\mu p = \mu(\Sigma a_i v_i) = \Sigma\mu(a_i)\mu v_i$$

$$(p \otimes 1 + 1 \otimes p)\mu = \Sigma(a_i \otimes 1)(v_i \otimes 1)\mu + \Sigma(1 \otimes a_i)(1 \otimes v_i)\mu = \Sigma(a_i \otimes 1 + 1 \otimes a_i)\mu v_i$$

$$Sp = \Sigma(Sa_i)Sv_i$$

$$pS = \Sigma a_i v_i S = -\Sigma a_i S v_i$$

$$\varepsilon p = \Sigma \varepsilon(a_i)\varepsilon v_i$$

$$p\varepsilon = 0$$

So if $a_i \in X_a(G)$ clearly $p \in P(G/K)$. Conversely, if $p \in P(G/K)$ then by the remark made at point 3) we have $a_i \in X_a(G)$ and we are done.

(1.9) PROPOSITION. Let G be an irreducible commutative algebraic K-group. Then, under the identification $\mathrm{Der}_K G \simeq \mathcal{O}(G) \otimes L(G)$ we have $P(G/K) = X_a(G) \otimes L(G)$.

Proof. Same computation made in (1.8) (with the "abusive" notational conventions from (1.1)).

Now we are in a postion to compute P(G/K) for a few special commutative G's.

EXAMPLE 0. Let $G = G_m^n = \mathrm{Spec}\, K[\chi_1, \dots, \chi_n]$ be a torus ($\mu\chi_i = \chi_i \otimes \chi_i$). Then $P(G/K) = 0$.

EXAMPLE 1. Let $G = G_a^n = \mathrm{Spec}\, K[\xi_1, \dots, \xi_n]$ be an algebraic vector group ($\mu\xi_i = \xi_i \otimes 1 + 1 \otimes \xi_i$). Then a K-basis of P(G/K) consists of the derivations $\xi_i \frac{\partial}{\partial \xi_j}$, $1 \leq i,j \leq n$ of $K[\xi_1, \dots, \xi_n]$.

EXAMPLE 2. Let $G = G_m \times G_a = \mathrm{Spec}\, K[\chi, \chi^{-1}, \xi]$ ($\mu\chi = \chi \otimes \chi$, $\mu\xi = \xi \otimes 1 + 1 \otimes \xi$). Then a K-basis of P(G/K) consists of the derivations $\xi \frac{\partial}{\partial \xi}$ and $\xi\chi \frac{\partial}{\partial \chi}$ of $K[\chi, \chi^{-1}, \xi]$.

EXAMPLE 3. Let $G = A \times G_a = A \times \mathrm{Spec}\, K[\xi]$ ($\mu\xi = \xi \otimes 1 + 1 \otimes \xi$) where A is the elliptic curve with $K(A) = K(x,y)$, $y^2 = x(x-1)(x-\lambda)$, $\lambda \in K$. Then a K-basis of P(G/K) consists of the derivations $\xi \frac{\partial}{\partial \xi}$ and $\xi y \frac{\partial}{\partial x}$ of $K(x,y,\xi)$.

(1.10) Let G be a K-group scheme. Then G(K) acts by adjoint representation on L(G): for any $g \in G(K)$ and $v \in L(G)$ we put $(\mathrm{Ad}g)(v) = L_g^{-1} v L_g$ where $L_g : \mathcal{O}_G \to L_{g*}\mathcal{O}_G$ is induced by left translation $L_g : G \to G$ defined by g. By (1.2) we dispose of a K-linear map

$$\nabla : P(G) \to \mathrm{Der}_k(L(G), L(G)), \quad p \mapsto \nabla_p$$

defined by $\nabla_p v = [p,v]$ (bracket in $\mathrm{Der}_k G$) for $p \in P(G)$, $v \in L(G)$. Then define a K-linear map

$$\ell\nabla : P(G) \to Z^1(G(K), L(G)), \quad p \mapsto \ell\nabla_p$$

by the formula

$$\ell\nabla_p g = L_g^{-1} p L_g - p \qquad \text{for } p \in P(G),\ g \in G(K)$$

The fact that $\ell\nabla_p g \in L(G)$ follows from the following computation:

$$\mu(L_g^{-1}pL_g - p) = \mu L_g^{-1}pL_g - \mu p = (L_g^{-1}\otimes 1)\mu pL_g - \mu p =$$
$$= (L_g^{-1}\otimes 1)(p\otimes 1 + 1\otimes p)\mu L_g - (p\otimes 1 + 1\otimes p)\mu =$$
$$= ((L_g^{-1}\otimes 1)(p\otimes 1 + 1\otimes p)(L_g\otimes 1) - (p\otimes 1 + 1\otimes p))\mu =$$
$$= ((L_g^{-1}pL_g - p)\otimes 1)\mu$$

It is easy to check that the maps ∇ and $\ell\nabla$ define an integrable logarithmic $P(G)$ - connection on $(G(K), L(G))$; in particular $L(G)$ has a structure of $D(G)$-module (in fact of Lie $D(G)$-algebra) such that for $p \in P(G)$, $v \in L(G)$ we have $pv = [p,v]$ (pv is v multiplied by p in the $D(G)$-module law; there is some danger of confusion here with the composition $p \circ v$ as endomorphisms of $\mathcal{O}_G$ so one should be careful to distinguish between the two possible meanings of the symbol pv: this is why we sometimes write $\nabla_p v$ instead of pv to denote $[p,v]$).

The above structure of $D(G)$-module on $L(G)$ will be sometimes called the "adjoint connection" on $L(G)$.

(1.11) PROPOSITION. Let G be an integral K-group scheme which is either affine or algebraic. Then the map

$$\partial \oplus \ell\nabla : P(G) \to \mathrm{Der}_k K \oplus Z^1(G(K),L(G))$$

is injective. Moreover for any $p \in P(G)$ and $g \in G(K)$ we have $\ell\nabla_p g = 0$ if and only if $g \in G_{K[p]}(K)$ (recall that this means by definition that $g : \mathrm{Spec}\, K \to G$ is a $K[p]$-map).

Proof. Assume $p \in P(G/K) = \mathrm{Ker}\,\partial$ is such that $\ell\nabla_p g = 0$ for all $g \in G(K)$. Then of course (1.8) holds for $L_{left}(G) := \{v \in \mathrm{Der}_K G;\ \mu v = (1\otimes v)\mu\}$; in particular picking a pseudo-basis $w_1, w_2, \ldots$ in $L_{left}(G)$ if G is affine (or simply a basis if G is algebraic) we may write $p = \Sigma a_i w_i$ with $a_i \in \mathcal{O}(G)$. Clearly $L_g^{-1}w_iL_g = w_i$ for all i so the relation $0 = \ell\nabla_p g = L_g^{-1}pL_g - p$ implies

$$\Sigma a_i w_i = \Sigma(L_g^{-1}a_i)w_i$$

hence by (1.8) $a_i = L_g a_i$ for all i and g which implies $a_i \in K$ for all i, hence that $p \in L_{left}(G)$ so $\mu p = (1\otimes p)\mu$. Applying $1\otimes\varepsilon$ to the latter equality we get $p = (1\otimes \varepsilon p)\mu$. But $\varepsilon p = p\varepsilon = 0$ so $p = 0$ and injectivity follows.

Next, for a fixed $g \in G(K)$ we have $\ell\nabla_p g = 0$ if and only if $L_g : G \to G$ is a $K[p]$-map which happens if and only if g is a $K[p]$-map from Spec K to G (because $L_g = \mu(1\otimes g)$ and $g = L_g\varepsilon$). Our proposition is proved.

(1.12) REMARK. Let G be an algebraic K-group and K_o be a field of definition (say $G \simeq G_o \otimes_{K_o} K$, G_o a K_o-group). For each $p \in \mathrm{Der}_{K_o} K$ let p^* denote the trivial lifting of $\partial(p)$ from K to $G \simeq G_o \otimes K$ (i.e. $p^* = 1 \otimes \partial(p)$). Then the map $G(K) \to L(G)$ defined by $g \mapsto \ell\nabla_{p^*} g$ coincides with Kolchin's logarithmic p-derivative $[K_1]$ p. 394; an argument for this can be found in $[B_1]$ p. 25.

In what follows we shall use the universal enveloping algebra of the Lie algebra of algebraic D-groups G in order to establish a D-analogue of the classical Lie correspondence between algebraic subgroups of G and their Lie algebras.

(1.13) Let L be a Lie D-algebra. Then the universal enveloping K-algebra U(L) has a natural structure of associative D-algebra induced by the D-module structure of the tensor algebra (0.5). So by (0.5) the dual $U(L)^{0}$ (which is a commutative algebra with convolution product) becomes a D-algebra.

Now for any K-group scheme G there is a natural K-algebra map

$$j_G : \mathcal{O}_{G,e} \to U(L(G))^{0}$$

defined as follows: for $f \in \mathcal{O}_{G,e}$, $j_G(f) = \hat{f}$ where for any $u \in U(L(G))$, $\hat{f}(u) = \varepsilon uf$ (here U(L(G)) acts on $\mathcal{O}_{G,e}$ by K-linear endomorphisms in the natural way). If in addition G has a structure of D-group scheme, then by (1.10) L(G) has an induced structure of Lie D-algebra and one easily checks that the map j_G above is a map of D-algebras.

(1.14) LEMMA. If char K = 0 and G is integral and either affine or algebraic then the map $j_g : \mathcal{O}_{G,e} \to U(L(G))^{0}$ in injective. In particular, if G is algebraic $\dim_K P(G/K) < \infty$.

REMARK. If char K > 0, P(G/K) is in general infinite dimensional even if G is algebraic, e.g. for $G = G_a$ (indeed apply (1.9) and the fact that $\dim_K X_a(G_a) = \infty$).

Proof. First it is easy to check the Lemma for G algebraic. Then if G is affine write $G = \lim\limits_{\leftarrow} G_i$ as in (1.6); we have $\mathcal{O}_{G,e} = \lim\limits_{\to} \mathcal{O}_{G_i,e}$ and a commutative diagram

$$\begin{array}{ccc}
\mathcal{O}_{G_i,e} & \xrightarrow{\;j_{G_i}\;} & U(L(G_i))^{o} \\
\downarrow & & \downarrow u_i \\
\mathcal{O}_{G,e} & \xrightarrow[\;j_G\;]{} & U(L(G))^{o}
\end{array}$$

where j_{G_i} and u_i are injective (for u_i use (1.8), 3)). This implies injectivity of j_G. The claim that $\dim_K P(G/K) < \infty$ for algebraic G follows form the fact that P(G/K) embeds into $\mathrm{Der}_K L(G)$.

(1.15) Recall from [DG] that if G is an (integral, to simplify) affine K-group scheme and H is a normal subgroup scheme of G, then the quotient G/H exists as an afine K-group scheme. It can be constructed as follows: let $G = \varprojlim G_i$ as in (1.6), let J_H be the ideal of H in $\mathcal{O}(G)$, $J_i = \mathcal{O}(G_i) \cap J_H$ (it is a Hopf ideal in $\mathcal{O}(G_i)$) and put $H_i = \mathrm{Spec}\,\mathcal{O}(G_i)/J_i$; then $G/H = \varprojlim G_i/H_i$. Assume now H is integral. By (1.8), 3) one can easily check that L(H) identifies with $\{v \in L(G);\ vJ_H \subset J_H\}$ and by (1.8) we have an exact sequence

$$0 \to L(H) \to L(G) \to L(G/H) \to 0$$

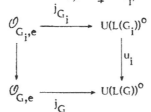

(1.16) PROPOSITION. Assume char $K = 0$ and H is an integral K-subgroup scheme of an integral D-group scheme G which is either affine or algebraic. The following are equivalent:

1) H is a D-subgroup scheme of G.

2) $L(H)$ is a D-submodule of $L(G)$.

In particular if G is algebraic then $[G,G]$ and $Z^{\circ}(G)$ are algebraic D-subgroups of G.

Proof. We have a commutative diagram

$$\begin{array}{ccc} \mathcal{O}_{G,e} & \xrightarrow{\ j_G\ } & U(L(G))^{\circ} \\ {\scriptstyle r}\downarrow & & \downarrow{\scriptstyle s} \\ \mathcal{O}_{H,e} & \xrightarrow[\ j_H\ ]{} & U(L(H))^{\circ} \end{array}$$

with r,s surjective and j_G, j_H injective (cf. (1.14) and (1.15)). If 2) holds then s is a D-algebra map hence its kernel is a D-ideal in $U(L(G))^{\circ}$. Since

$$\operatorname{Ker} r = j_G^{-1}(\operatorname{Ker} s)$$

we get that $\operatorname{Ker} r$ is a D-ideal which implies that H is a D-subscheme, hence a D-subgroup scheme of G. Conversely, if the latter holds the ideal sheaf J_H of H in $\mathcal{O}_G$ is a D-ideal sheaf in $\mathcal{O}_G$ hence if $v \in L(H)$, identifying v with an element (still denoted by v) in $L(G)$ such that $vJ_H \subset J_H$, we have (for any $p \in P$) $(\nabla_p v)J_H = (p \circ v)(J_H) + (v \circ p)(J_H) \subset J_H$, so $\nabla_p v \in L(H)$. The last assertion of the lemma follows from the fact that $L([G,G]) = [L(G),L(G)]$ and $L(Z^{\circ}(G)) = Z(L(G))$ (= center of $L(G)$) are clearly D-submodules of $L(G)$.

(1.17) LEMMA. Assume char $K = 0$, G is an integral D-group scheme which is either algebraic or affine and H is a normal (nonnecessary irreducible) K-subgroup scheme. The following are equivalent:

1) H is a D-subgroup scheme

2) There is a structure of D-group scheme on G/H such that the projection $G \to G/H$ is a D-map.

(Clearly the D-group scheme structure on G/H in 2) must be unique !).

Proof. 2)$\Rightarrow$1) is clear by (1.14), 7).

1)$\Rightarrow$2) By (0.14), 5) it is sufficient to check that $K(G/H)$ is a D-subfield of $K(G)$. If G is affine let $G = \varprojlim G_i, G/H = \varprojlim G_i/H_i$ as in (1.15). If G is algebraic, put $G_i = G$, $H_i = H$ for all i. Letting $\bar{\mu}_i : G_i \times H_i \to G_i$ be deduced from the multiplication $\mu_i : G_i \times G_i \to G_i$ we have

$$K(G/H) = \varinjlim K(G_i/H_i) = \varinjlim \{ f \in K(G_i);\ \bar{\mu}_i f = f \otimes 1 \}$$

where we still denoted (as usual) by $\bar{\mu}_i$ the induced map $\mathcal{O}_{G_i} \to \bar{\mu}_{i*}\mathcal{O}_{G_i \times H_i}$. Now if $f \in K(G_i)$ is such that $\bar{\mu}_i f = f \otimes 1$ and if $p \in P$ then $\bar{\mu} f = f \otimes 1$ ($\bar{\mu} : G \times H \to G$ having the obvious meaning) so

$$\bar{\mu} pf = (p \otimes 1 + 1 \otimes p)\bar{\mu} f = (p \otimes 1 + 1 \otimes p)(f \otimes 1) = pf \otimes 1$$

But $pf \in K(G_j)$ for some j hence $pf \in K(G/H)$ and we are done.

(1.18) PROPOSITION. Assume char $K = 0$, G is an irreducible algebraic D-group and H any algebraic D-subgroup of G. Then the homogenous space G/H has a natural structure of D-variety such that the projection map $G \to G/H$ is a D-map.

Proof. Same argument as in (1.17).

(1.19) PROPOSITION. Let $G = G_1 \times G_2$ be a product of K-group schemes. Then there is a natural Lie K/k-algebra map

$$s : P(G_1) \times_{\mathrm{Der}_k K} P(G_2) \to P(G)$$

and a natural K-linear map "over" $\mathrm{Der}_k K$:

$$\Pi : P(G) \to P(G_1) \times_{\mathrm{Der}_k K} P(G_2)$$

such that Πs = identity. In particular for any $p \in P(G)$ we have $p - s\Pi p \in P(G/K)$.

Proof. Let $\Pi_i : G \to G_i$ be the canonical projections (and denote as usual also by Π_i the induced maps $\mathcal{O}_{G_i} \to \Pi_{i*}\mathcal{O}_G$). Then for $p \in P(G)$ define $\Pi(p) = ((1 \otimes \varepsilon_2)p\Pi_1, (1 \otimes \varepsilon_1)p\Pi_2)$ and $s(p_1, p_2) = p_1 \otimes 1 + 1 \otimes p_2$ where of course ε_i is the co-unit of G_i. Then one checks easily all claims of the proposition.

The next proposition shows that the conditions defining the elements of P(G), cf. (1.1), can be weakened in case G is algebraic over K. This proposition will play an important role later.

(1.20) PROPOSITION. Let char $K = 0$ and G be an irreducible algebraic K-group. Assume $p \in \mathrm{Der}_k K(G)$ is such that

1) $pK \subset K$

2) $\mu p = (p \otimes 1 + 1 \otimes p)\mu : K(G) \to K(G \times G)$

3) $Sp = pS : K(G) \to K(G)$.

Then $p \in P(G)$.

Proof. Start with the remark that if K_1/K is a K[p]-field extension (0.10) then p extends to some derivation $p_1 \in \mathrm{Der}_k K_1(G)$ satisfying the corresponding properties 1)-3) from the statement of the proposition. Since the projection $G \otimes_K K_1 \to G$ is faithfully flat it is sufficient by (0.14), 5) to check that $p_1 \in P(G \otimes_K K_1)$. So we may assume that K equipped with the derivation $\partial(p) \in \mathrm{Der}_k K$ is K[p]-algebraically closed. By (0.14) again, the locus $(p)_\infty = \{x \in G; p(\mathcal{O}_{G,x}) \not\subset \mathcal{O}_{G,x}\}$ is a divisor on G. Put $U = G \setminus (p)_\infty$; it is a K[p]-variety.

The set of K[p]-K-points $U_{K[p]}(K)$ of U is dense in U(K). Take $g \in U_{K[p]}(K)$ and consider the corresponding left translation

$$L_g : G \xrightarrow{g \times 1} G \times G \xrightarrow{\mu} G$$

Its restriction

$$(g \times 1)^{-1}(\mu^{-1}(U) \cap (U \times U)) \xrightarrow{g \times 1} \mu^{-1}(U) \cap (U \times U) \xrightarrow{\mu} U$$

is a K[p]-morphism (where note that the source scheme is an open subset of U which is nonempty because $U \cap g^{-1}U \neq \emptyset$). Consequently the induced morphism $L_g : K(G) \to K(G)$ commutes with $p : K(G) \to K(G)$ this showing that $(p)_\infty$ is L_g-invariant for all $g \in U_{K[p]}(K)$. By density this happens for all $g \in G(K)$ so $(p)_\infty$ must be empty, hence $p\mathcal{O}_G \subset \mathcal{O}_G$. Finally, take any $g \in G_{K[p]}(K)$; since Sp = pS, the morphism

$$\varepsilon : \operatorname{Spec} K \xrightarrow{g \times g^{-1}} G \times G \xrightarrow{\mu} G$$

is a K[p]-morphism and we are done.

In what follows, by an isogeny we mean a surjective morphism $G' \to G$ of irreducible algebraic groups having finite kernel.

(1.21) PROPOSITION. Let $f : G' \to G$ be a separable isogeny. There exists a canonical injective "lifting" map of Lie K/k-algebras $f^* : P(G) \to P(G')$. Moreover, if G and G' are commutative and char K = 0 then f^* is an isomorphism.

Proof. Let μ', S', ε' be the structure maps of G'. Since f is étale, any $p \in \operatorname{Der}_k G$ lifts uniquely to some $p' \in \operatorname{Der}_k G'$. Now if $p \in P(G)$, $\mu' p' - (p' \otimes 1 + 1 \otimes p')\mu'$ is a μ'-K-derivation $K(G') \to K(G' \times G')$ vanishing on K(G); by separability this derivation must vanish. Similarily we get $p'S' = S'p'$ and $p'\varepsilon' = \varepsilon' p'$. Of course we put $f^*(p) = p'$ which is well defined by the preceeding remarks and clearly is an injective Lie K/k-algebra map from P(G) to P(G').

Now assume char K = 0 and G,G' commutative. Then there is an isogeny $g : G \to G'$ such that $gf : G' \to G \to G'$ is the multiplication by some integer $N \geq 1$, which we call λ_N. We get injective maps

$$\lambda_N^* : P(G') \xrightarrow{g^*} P(G) \xrightarrow{f^*} P(G')$$

We claim that λ_N^* is the identity (this will close our proof). This follows by noting that λ_N is a map of D(G')-varieties hence for any $p' \in P(G')$ we have a commutative diagram

$$\begin{array}{ccc} K(G') & \xrightarrow{\lambda_N} & K(G') \\ {\scriptstyle p'}\downarrow & & \downarrow{\scriptstyle p'} \\ K(G') & \xrightarrow[\lambda_N]{} & K(G') \end{array}$$

showing that the canonical lifting of p' via λ_N is p' itself. Our proposition is proved.

(1.22) PROPOSITION. Let G be an irreducible commutative algebraic D-group. Then the following hold:

1) Any torsion point in G(K) of order prime to char K is a D-point (in other words if N is an integer prime to char K, $N\text{-Tors}(G(K)) \subset G_D(K)$).

2) Any torus and any abelian variety contained in G is an algebraic D-subgroup of G.

3) Any non-linear irreducible algebraic subgroup of G contains a non-trivial irreducible algebraic D-subgroup of G.

Proof. 1) Let $g \in N\text{-Tors}(G(K))$. Then multiplication by N (call it $\lambda_N : G \to G$) is a separable isogeny and $\lambda_N^{-1}(0)$ is a D-subscheme of G, union of reduced points. By (0.14), each of these points is a D-point, in particular so is g.

2) Let H be either a torus or an abelian variety contained in G. Then the torsion points of H whose order is prime to char K is dense in H; so on any affine open set of G the defining ideal $\mathfrak{J}$ of H in G is an intersection of D-ideals so $\mathfrak{J}$ itself is a D-ideal and we are done.

3) Any non-linear irreducible algebraic subgroup H of G has infinitely many torsion points of order a power of a prime ℓ where ℓ does not divide char K. Let M be the Zariski closure in H of the group of these torsion points. Then M is an algebraic D-subgroup of H and so will be its connected component.

(1.23) Let's make some general remarks on the link between P(G) and the automorphism functor of G (a deeper discussion will be done in Chapter 4).

For any K-group scheme G we have the automorphism functor:

$$\underline{\mathrm{Aut}}\, G : \{K\text{-schemes}\} \to \{\text{groups}\}$$

defined by $(\underline{\mathrm{Aut}}\, G)(S) = \mathrm{Aut}_{S\text{-gr.sch}}(G \times S)$ (in particular $(\underline{\mathrm{Aut}}\, G)(K) = \mathrm{Aut}\, G$). This functor need not be representable even if char K = 0 and if G is an algebraic group (cf. [BS], e.g. if $G = G_a \times G_m$). It is representable however if G is algebraic reductive (and char $K \geq 0$) cf. [GD]. Recall from [GD] that for any contravariant functor $\underline{A} : \{K\text{-scheme}\} \to \{\text{groups}\}$ one can define the "Lie algebra"

$$L(\underline{A}) := \mathrm{Ker}\,(A(\Pi) : \underline{A}(\mathrm{Spec}\, K[\varepsilon]) \to \underline{A}(\mathrm{Spec}\, K))$$

where $K[\varepsilon] = K + K\varepsilon$, $\varepsilon^2 = 0$ and the map $\Pi : K[\varepsilon] \to K$ is defined by $\varepsilon \to 0$; $L(\underline{A})$ is a group but not a genuine Lie K-algebra in general. In our specific case however, $L(\underline{\mathrm{Aut}}\, G)$ is easily seen to identify with $P(G/K) = \mathrm{Ker}(\partial : P(G) \to \mathrm{Der}_k K)$, in particular it is a genuine Lie K-algebra; the identification is given of course by the map $p \to id + \varepsilon p$ $(p \in P(G/K))$.

There are remarkable cases when the restriction of $\underline{\mathrm{Aut}}\, G$ to {reduced K-schemes} is representable by a locally algebraic K group, call it in this case Aut G (there is no danger of confusion with the other meaning of Aut G which is $(\underline{\mathrm{Aut}}\, G)(K)$!). This is the case if char K = 0 and G is linear algebraic [BS] (or even char K = 0 and G is algebraic, non-necessarily linear, cf. (IV.2) below).

If the above representability properties holds then there is an induced Lie K-algebra map

$$\lambda = \lambda_G : L(\text{Aut}\, G) \to L(\underline{\text{Aut}}\, G)$$

which is not an isomorphism in general and which will be studied in (IV.2).

(1.24) For any K-group scheme the group $(\underline{\text{Aut}}\, G)(K) = \text{Aut}\, G$ acts on $\text{Der}_k G$ by the formula $(\sigma,p) \to \sigma^{-1} p \sigma$ for any $\sigma \in \text{Aut}\, G$, $p \in \text{Der}_k G$ (as usual we still denote by σ the map $\mathcal{O}_G \to \sigma_* \mathcal{O}_G$ induced by σ). Under this action P(G) and P(G/K) are globally invariant. Let us call

$$\nabla : P(G) \to \text{Der}_k(P(G/K), P(G/K)), \quad p \mapsto \nabla_p$$

the K-linear map defined by $\nabla_p v = [p,v]$ (bracket taken in P(G)) for $p \in P(G)$, $v \in P(G/K)$ and define the K-linear map

$$\ell\nabla : P(G) \to Z^1(\text{Aut}\, G, P(G/K)), \quad p \mapsto \ell\nabla_p$$

by the formula $\ell\nabla_p \sigma = \sigma^{-1} p \sigma - p$ for $p \in P(G)$, $\sigma \in \text{Aut}\, G$. One easily checks that $(\nabla, \ell\nabla)$ above define an integrable logarithmic P(G)-connection on (Aut G, P(G/K)).

(1.25) In way of our study of P(G) for commutative G we shall be lead to consider one more example of logarithmic $\text{Der}_k K$-connection $(\nabla, \ell\nabla)$ whose "additive part" ∇ is the Gauss--Manin connection [KO] on the first de Rham cohomology space of abelian varieties while the "multiplicative part" $\ell\nabla$ will be defined and used in Chapter 3.

(1.26) We close this section by noting that if char K = 0 and if K_0 is a field of definition for a K-group scheme G (i.e. if we have a K-group scheme isomorphism $\sigma : G \simeq G_0 \otimes_{K_0} K$, G_0 a K_0-group scheme) then there is a Lie K/k-algebra map

$$\text{Der}(K/K_0) \to P(G), \quad p \mapsto p^*$$

where $p^* = \sigma(1 \otimes p)\sigma^{-1}$ is the trivial lifting of p to G (it depends on σ !). We shall often identify $\text{Der}(K/K_0)$ with a Lie K/k-subalgebra of P(G) (after having fixed σ !). It is easy to check that $K^{P(G)}$ is contained in any algebraically closed field of definition for G containing k. If $K^{P(G)}$ is a field of definition for G (which is not always the case, cf. Chapter 3 below !) then we have a split exact sequence of Lie K/k-algebras

$$0 \to P(G/K) \to P(G) \xrightarrow{\partial} \text{Der}(K/K^{P(G)}) \to 0$$

In particular P(G) is the semidirect product of Ker ∂ and Im ∂ .

EXAMPLES

Let's consider once again the Examples 1, 2, 3 after (1.9), where in Example 3 we assume

$\lambda \in k$.

For any $p \in \mathrm{Der}_k K$ let p^* denote the unique lifting of p to $K[\xi_1,\ldots,\xi_n]$, $K[\chi,\chi^{-1},\xi]$, $K(x,y,\xi)$ for which $p^*\xi_i = 0$, $p^*\chi = 0$, $p^*x = 0$, $p^*y = 0$, $p^*\xi = 0$. By the discussion above $p^* \in P(G)$ in each of the three cases. So we may describe the space P(G) in each of the three cases as follows.

Indeed for $G = G_a^n$ (Example 1), P(G) consists of all derivations of the form:

$$\tilde{p} = p^* + \Sigma a_{ij}\xi_i \frac{\partial}{\partial \xi_j}\ , \quad a_{ij} \in K$$

Next for $G = G_m \times G_a$ (Example 2) P(G) consists of all derivations of the form

$$\tilde{p} = p^* + a\xi\chi\frac{\partial}{\partial \chi} + b\xi\frac{\partial}{\partial \xi}\ , \quad a,b \in K$$

Finally for $G = A \times G_a$ (Example 3) P(G) consists of all derivations of the form

$$\tilde{p} = p^* + a\xi y\frac{\partial}{\partial x} + b\xi\frac{\partial}{\partial \xi}\ , \quad a,b \in K$$

In connection with fields of definition, the following result proved in $[B_4]$, $[B_5]$ will be needed later:

(1.27) THEOREM. Let char K = 0 and G be an irreducible algebraic K-group (respectively L a finite dimensional Lie K-algebra). Then the set of all algebraically closed fields of definition for G (respectively for L) between k and K has a minimum element K_G (respectively K_L). If in addition G is affine then

$$K_G = K_{L(U)}$$

where U is the unipotent radical of G.

2. Prolongations and embeddings

In this section we assume for the sake of simplicity (and due to the fact that this is suficient for our applications in Chapter V) that P has a finite commuting K-basis $\{p_1,\ldots,p_m\}$; so a K-basis for D will consists of "commutative" monomials in the variables p_i. We also assume char K = 0.

(2.1) Consider the forgetful functor

$$\{\text{reduced D-schemes}\} \to \{\text{reduced K-schemes}\}\ , \quad V \mapsto V^!$$

One can construct a right adjoint $X \mapsto X^\infty$ to this functor using Johnson's "prolongation" procedure [J] (see also the concept of "produced scheme" from [NW]). So, for any reduced K-scheme X and any reduced D-scheme V we will have a natural bijection

$$\mathrm{Hom}_{K\text{-sch}}(V^!,X) \simeq \mathrm{Hom}_{D\text{-sch}}(V,X^\infty)$$

Note we could have done this construction in the non-reduced case too; we restricted ourselves to the reduced case only because this is sufficient for our applications.

Let's "recall" one of the possible constructions of $X \mapsto X^{\infty}$. We construct a sequence $A_n (n \geq -1)$ of sheaves of $\mathcal{O}_X$-algebras on X equipped with $\mathcal{O}_X$-algebra maps

$$f_n : A_n \to A_{n+1}$$

and with f_n-k-derivations $(1 \leq i \leq m)$

$$d_n^i : A_n \to A_{n+1}$$

inductively starting with $A_{-1} = K$, $A_o = \mathcal{O}_X$, f_{-1} = natural inclusion $K \subset \mathcal{O}_X$, $d_{-1}^i = \partial(p_i) : K \to K \subset \mathcal{O}_X$ and then letting

$$A_{n+1} = S(\Omega^1_{A_n/k} \otimes_K P)/J_n$$

where J_n is the sheaf of ideals (in the symmetric A_n-algebra of $\Omega^1_{A_n/k} \otimes_K P$) generated by elements of the form

(2.1.1) $\quad d_{n-1}^i a - (d(f_{n-1}(a))) \otimes p_i$, $\quad a \in A_{n-1}$ and

(2.1.2) $\quad d(d_{n-1}^j a) \otimes p_i - d(d_{n-1}^i a) \otimes p_j$, $\quad a \in A_{n-1}$

where $d : A_n \to \Omega^1_{A_n/k}$ is the usual differential. Moreover we let f_n be induced by the natural inclusion map $A_n \to S(\Omega^1_{A_n/k} \otimes P)$ and d_n^i be induced by the map

$$A_n \to \Omega^1_{A_n/k} \otimes P, \quad b \mapsto (db) \otimes p_i$$

We put then $A^{\infty} = (\varinjlim A_n)_{red}$, $d^i = (\varinjlim d_n^i)_{red}$ and $X^{\infty} = \operatorname{Spec} A^{\infty}$. Then $d^1, \ldots, d^m$ are pairwise commuting derivations on A^{∞} hence define a structure of D-scheme on X^{∞}. One easily checks that the functor $X \mapsto X^{\infty}$ is the one we are looking for.

If $\Pi : (X^{\infty})^! \to X$ is the natural map given by adjunction, then Π is clearly affine and for any open set $U \subset X$, $U^{\infty} = \Pi^{-1}(U)$. Moreover, if X is of finite type over K then X^{∞} of D-finite type.

As the functor $X \mapsto X^{\infty}$ has a left adjoint, it commutes with direct products so it induces in a natural way a functor

$$\{\text{reduced K-group schemes}\} \to \{\text{reduced D-group schemes}\}, \quad G \mapsto G^{\infty}$$

The functor above is again a right adjoint for the forgetful functor. Clearly, if G is commutative so is G^{∞}. Of course, if G is algebraic, G^{∞} will be not so (except if G is finite). By adjunction we have for any algebraic K-group H and for any algebraic D-group G morphisms

$$(H^{\infty})^! \to H$$

$$G \to (G^!)^\infty$$

which we are briefly investigating in what follows.

(2.2.) LEMMA. For any field extension K_1/K the map $(H^\infty)^!(K_1) \to H(K_1)$ is surjective.

Proof. Any field extension K_1 of K admits a D-field structure and we are immediately done by adjunction.

(2.3) PROPOSITION. Assume G is an irreducible algebraic D-group. Then:

1) There exists a naturally associated D-cocycle

$$\ell\Delta : G^{!\infty} \to (L(G)^m)^\infty = (L(G)^\infty)^m$$

such that the morphism $G \to G^{!\infty}$ deduced from adjunction induces an isomorphism of D-group schemes between G and the reduced kernel $(\ker \ell\Delta)_{red}$ of $\ell\Delta$. (Here $L(G) = L(G^!)$ is of course identified with its associated algebraic vector group and the action of $G^{!\infty}$ on $(L(G)^m)^\infty$ is induced by functoriality from the adjoint action of G on each factor of $L(G)^m$).

2) There is a bijection $PHS_D(G/K) \simeq L(G)^{int}/G(K)$ where $L(G)^{int} = \{(v_1, \ldots, v_m) \in L(G)^m;\ \nabla_{p_i} v_j - \nabla_{p_j} v_i + [v_i, v_j] = 0$ for all $i,j\}$ (cf. notations in (1.10)) and G(K) acts on $L(G)^{int}$ by the "Loewy-type" $[C_1]$ formula:

$$(g,(v_i)_i) \to (L_g^{-1} v_i L_g + \ell\nabla_{p_i} g)_i \qquad (g \in G(K))$$

(where $\ell\nabla_{p_i} g = L_g^{-1} p_i L_g - p_i$, cf. (1.10)). In particular, if G is commutative $PHS_D(G/K)$ has a natural structure of abelian group.

3) There is an exact sequence of pointed sets (which in case G is commutative is an exact sequence of abelian groups):

$$1 \to G_D(K) \to G(K) \xrightarrow{\ell\Delta} L(G)^{int} \to PHS_D(G/K) \to 1$$

In particular, if K is D-algebraically closed, $\ell\Delta$ above is surjective.

REMARK. The last surjectivity assertion is a generalisation of Kolchin's surjectivity theorem for the logarithmic derivative (cf. $[K_1]$ pp. 420-421, or [NW]); it reduces to it in case G is split (cf. (3.11) below for the definition of "split") but our proof here is different from those in $[K_1]$, [NW].

Proof. 1) To define $\ell\Delta$ it is sufficient by (0.9) to define for any reduced D-algebra A, cocycles

$$\ell\Delta_A : (G^{!\infty})_D(A) = G^!(A^!) \to ((L(G)^m)^\infty)_D(A) = L(G)^m \otimes A$$

behaving functorially in A. We shall define $\ell\Delta_A$ on each component to be given by the map

$$(*) \qquad g \to L_g^{-1} p_i L_g - p_i \qquad (g \in G^!(A^!))$$

where p_i denotes here the derivation of $\mathcal{O}_{G\otimes A}$ corresponding to $p_i \in P$ while $L_g : \mathcal{O}_{G\otimes A} \to \mathcal{O}_{G\otimes A}$ is deduced from the left translation by g. Then $L_g^{-1} p_i L_g - p_i \in L(G) \otimes A$ because

$$\mathrm{Der}_A(\mathcal{O}_{G\otimes A}) = \mathrm{Hom}_{G\otimes A}(\Omega^1_{G\otimes A/A}, \mathcal{O}_{G\otimes A}) = (\mathrm{Der}_K G) \otimes_K A$$

and because one can transpose mutatis mutandis the computation made in (1.10) (note that for $A = K$ our map (*) is precisely the map $\ell\nabla_P$ from (1.10)). To prove that $G \simeq (\mathrm{Ker}\, \ell\Delta)_{red}$ it is sufficient to check that we have for any reduced D-algebra A an exact sequence of pointed sets

$$1 \to \mathrm{Hom}_{D\text{-sch}}(\mathrm{Spec}\, A, G) \to \mathrm{Hom}_{K\text{-sch}}(\mathrm{Spec}\, A, G^!) \xrightarrow{\ \ell\Delta_A\ } L(G)^m \otimes A$$

This can be done by adapting mutatis mutandis the last part of the proof of (1.11) (which deals with the case A = K!). This closes the proof of 1).

To prove 2) we may identify (in a non-canonical way !) the underlying varieties of all D-principal homogenous spaces for G with G itself (since K is algebraically closed !) so to give a D-principal homogenous space for G is equivalent to giving commuting derivations $d_1, \ldots, d_m \in \mathrm{Der}_k G$ satisfying the formula

$$\mu d_i = (d_i \otimes 1 + 1 \otimes p_i)\mu$$

(where $d_i \otimes 1 + 1 \otimes p_i$ is of course the derivation on G x G induced by d_i on the first factor and by p_i on the second).

The above condition is equivalent to

$$\mu(d_i - p_i) = ((d_i - p_i) \otimes 1)\mu$$

hence with $d_i - p_i \in L(G)$ (once again we still denote by p_i the derivation in $\mathrm{Der}_k G$ corresponding to p_i). Now two m-uples $(d_1, \ldots, d_m)$, $(d'_1, \ldots, d'_m)$ give isomorphic D-principal homogenous spaces if and only if there is an isomorphism of K-varieties $\phi : G \to G$ making the following diagram commutative

$$\begin{array}{ccc} G \times G & \xrightarrow{\ \mu\ } & G \\ {\scriptstyle \phi \times 1}\downarrow & & \downarrow{\scriptstyle \phi} \\ G \times G & \xrightarrow[\ \mu\ ]{} & G \end{array}$$

such that $d'_i = \phi^{-1} d_i \phi$ for all i. The above diagram shows that ϕ must be the left translation L_g ($g = \phi(e) \in G(K)$) and so the latter relation becomes

$$L_g^{-1}(d_i - p_i)L_g + \ell\nabla_{p_i} g = d'_i - p_i$$

which proves 2).

Assertion 3) follows directly from 1) and 2).

3. Local finitness and splitting

Here we come to one of our main concepts (and tools) namely to "local finitness" and its relation with "splitting" and "fields of definition"; this is part of our ideology in $[B_1]$. We start with some definitions (where we assume K algebraically closed as usual and of arbitrary characteristic while P is any Lie K/k-algebra and D = K[P]).

(3.1) A D-module V is called finite dimensional if $\dim_K V < \infty$. A D-module is called locally finite if it is the union (equivalently the sum) of its finite dimensional D-submodules. Note that if V and W are locally finite D-modules then $V \oplus W$ and $V \otimes W$ are locally finite but $\mathrm{Hom}_K(V,W)$ won't be so in general and in fact not even $V^o = \mathrm{Hom}_K(V,K)$ will be locally finite in general. A D-module V is split [NW][B_1] if V has a K-basis contained in V^D, equivalently if the natural K-linear map

$$K \otimes_{K^D} V^D \to V$$

(which is always injective) is also surjective. Of course, if char K > 0 then V is split if and only if P acts trivially on V so the concept of "splitting" will be interesting only in characteristic zero. Any D-submodule of a split D-module is easily seen to be split. Moreover, any split D-Module is locally finite. Conversely, we have the following result essentially proved in $[B_1]$, p. 85 (confer also with [T] for a generalisation and also to $[B_3]$):

(3.2) LEMMA. Assume char K = 0 and V is locally finite D-module having at most a countable dimension over K. Then there exists a D-field extension K_1/K such that $K_1^D = K^D$ and such that $K_1 \otimes_K V$ is a split D_1-module ($D_1 = K_1 \otimes_K D$ as in (0.10)). Moreover if we assume K is D-algebraically closed we may take $K_1 = K$.

Proof. For the sake of completness we sketch the argument. It is clear that we may assume $N = \dim_K V < \infty$. Then the D-module structure of V induces a D-variety structure on GL_N by just fixing a basis $v_1, \ldots, v_N$ in V: if $pv_i = \Sigma a_{ij}(p)e_j$ with $a_{ij}(p) \in K$ and if $GL_N = \mathrm{Spec}\, K[X_{ij}]_d$, $d = \det(X_{ij})$ then we let

$$pX_{ij} = \sum_k a_{ik}(p)X_{kj} \qquad \text{for all } i,j,p$$

Now let g be a maximal element of $(GL_N)_D$ and put $K_1 = K(g)$ (residue field at p; it is a D-field extension of K). By (0.14), we have $K_1^D = K^D$ (because K^D is algebraically closed). Viewing g as a K_1-point of GL_N, say $g = (g_{ij})$, $g_{ij} \in K_1$ it is easy to see that the elements $f_1, \ldots, f_N \in K_1 \otimes_K V$ defined by $e_i = \Sigma g_{ij} f_j$ form a K_1-basis of $K_1 \otimes V$ belonging to $(K_1 \otimes V)^{D_1}$.

(3.3) A D-algebra A (respectively a Lie D-algebra L, or a Hopf D-algebra H) is called locally finite if it is so as a D-module; A (respectively L,H) is called split [NW] if there exists a D-algebra isomorphism $A \simeq K \otimes_{K^D} A_0$ where A_0 is a K^D-algebra and $K \otimes_{K^D} A_0$ is viewed as a D-algebra with P acting via ∂ on $K \otimes 1$ and trivially on $1 \otimes A_0$ (similar definition for L,H). Here is a remarkable (trivial) remark:

(3.4) LEMMA. Let A be a D-algebra (respectively L a Lie D-algebra of H a Hopf D-algebra). Then A (respectively L,H) is split as a D-algebra (respectively L is split as a Lie D-algebra or H is split as a Hopf D-algebra) if and only if A (respectively L, H) is split as a D-module.

Proof. Let's a give the proof for H, for instance. Assume H is split as a D-module. Then it is easy to check that H^D is a K^D-subalgebra of H invariant under antipode and mapped by comultiplication into $H^D \otimes_{K^D} H^D = (H \otimes_K H)^D$ so that H^D has a naturally induced structure of Hopf K^D-algebra. Moreover the natural map $K \otimes_{K^D} H^D \to H$ is clearly a Hopf D-algebra isomorphism, hence H is split as a Hopf D-algebra. The converse is obvious.

(3.5) Let X be a D-scheme; an open set U of X will be called D-invariant if there is a closed D-subscheme whose support is $X \setminus U$. Now X will be called locally finite if every point in X has an affine D-invariant open neighbourhood whose coordinate algebra is locally finite. Note that if X is locally finite and quasi-compact then the D-algebra $\mathcal{O}(X)$ is locally finite; indeed if $(U_i)_i$ is a finite covering with affine open sets such that $\mathcal{O}(U_i)$ is locally finite then $\mathcal{O}(X)$ appears as a D-submodule of $\oplus \mathcal{O}(U_i)$ which is locally finite hence it is locally finite. In particular this shows that if X is an affine D-scheme then X is locally finite if and only if $\mathcal{O}(X)$ is locally finite.

(3.6) Recall $[B_1]$ that a D-scheme X is called split if there is an isomorphism of D-schemes $X \simeq K \otimes_{K^D} X_0$ where X_0 is a K^D-scheme and $K \otimes_{K^D} X_0$ is viewed as a D-scheme with P acting on $K \otimes 1$ via ∂ and acting trivially on $\mathcal{O}_{X_0}$. Recall some basic facts about split D-schemes:

0) Any split D-scheme is locally finite.

1) If X_0 and Y_0 are K^D-scheme and if $X = K \otimes_{K^D} X_0$, $Y = K \otimes_{K^D} Y_0$ are viewed with their natural structures of split D-schemes then any D-morphism $\phi : X \to Y$ has the form $\phi = K \otimes \phi_0$ for some morphism $\phi_0 : X_0 \to Y_0$. In particular $X_D(K)$ identifies with $X_0(K^D)$.

2) Assume the D-scheme X is covered by D-invariant open subsets U_i each of which is split; then X itself is split.

3) Any closed D-subscheme of a split D-scheme is split.

4) Any D-invariant open D-subscheme of a split D-scheme is split.

The proof of the above assertions can be done as follows: first prove 1) for affine schemes using the fact that D-algebra maps must carry P-constants into P-constants. Then prove 3) in the affine case using the fact that any D-submodule of a split D-module is split.

Now 4) in the affine case follows from 1) and 3) in the affine case. Next prove 2) by first noting that the affine case of 4) implies that for any set of indices $J = \{j_1, \dots, j_N\}$, $U_J = U_{j_1} \cap \dots \cap U_{j_N}$ is affine and split (use separation of X) hence by 1) in the affine case again we may "glue together" the splittings of the U_i's. Finally one uses 2) to reduce 1), 3), 4) to the affine case.

(3.7) PROPOSITION. Assume char K = 0 and X is a locally finite D-variety. Then:

1) Any D-invariant affine open set of X is locally finite.

2) There exists a D-field extension K_1/K with $K_1^D = K^D$ such that $K_1 \otimes_K X$ is a split D_1-variety, where $D_1 = K_1 \otimes_K D$. Moreover, if K is D-algebraically closed then X itself is split.

Proof. First prove 2). Cover X by affine D-invariant open subsets U_i such that $\mathcal{O}(U_i)$ is locally finite. By (3.2) we may find an extension of D-fields K_1/K with $K_1^D = K^D$ (in case K is D-alebraically closed we may take $K_1 = K$) such that each $K_1 \otimes_K U_i$ is split (one simply applies (3.2) to $V = \oplus V_i$ where V_i is a finite dimensional D-submodule of $\mathcal{O}(U_i)$ generating $\mathcal{O}(U_i)$ as a K-algebra). Clearly $K_1 \otimes U_i$ are D_1-invariant open subsets of $K_1 \otimes X$ and we conclude by assertion 2) in (3.6) that $K_1 \otimes X$ is split.

To prove 1) let K_1/K be as in 2) and let $U \subset X$ be an affine open D-invariant subset. Then $K_1 \otimes U$ is D_1-invariant. By 4) in (3.6) $K_1 \otimes U$ is split, in particular $\mathcal{O}(K_1 \otimes U)$ is locally finite as a D_1-module. It follows that $\mathcal{O}(U)$ is locally finite as a D-module as well. Our proposition is proved.

(3.8) REMARK. Let X be a D-variety, char K = 0. It may happen that there exists a D-field extension K_1/K with $K_1^D = K^D$ such that $K_1 \otimes_K X$ is a split D_1-variety ($D_1 = K_1 \otimes_K D$) but X is not locally finite: take for instance X to be an elliptic curve over K viewed as a K[p]-variety where p acts on X via a non-zero global vector field v on X. In this case X contains no non-empty K[p]-invariant open set (because v is nowhere vanishing) but X splits over a "strongly normal" extension of K by $[B_1]$ p. 46.

(3.9) Let G be a K-group scheme. We denote by P(G,fin) the set of all $p \in P(G)$ such that G is a locally finite K[p]-scheme (if G is affine this is equivalent to saying that $\mathcal{O}(G)$ is a locally finite K[p]-module, cf. (3.5)). We also put

$$P(G/K,\mathrm{fin}) = P(G/K) \cap P(G,\mathrm{fin})$$

We will prove later (cf. (IV. 4.1)) that if G is algebraic (irreducible) and char K = 0 then P(G,fin) is a Lie K/k-subalgebra of P(G). But note that if char K = q > 0 then P(G,fin) (which coincides with P(G/K,fin)) need not be a linear subspace of P(G) even if G is algebraic. Let's give such an example in what follows.

First it is easy to check that for q = 2,3,5 the following formula holds in the polynomial ring $A = \mathbf{F}_q[\xi]$:

$$(*) \qquad \left((\xi - \xi^q)\frac{\partial}{\partial \xi} + \xi \cdot 1_A\right)^{q-1}(\xi) = \xi$$

(Presumably this formula holds for any prime q!).

Assuming formula (*) holds for a given prime q (e.g. assuming $q \in \{2,3,5\}$) let $G = G_a \times G_m = \mathrm{Spec}[\xi,\chi,\chi^{-1}]$, $\mu\xi = \xi \otimes 1 + 1 \otimes \xi$, $\mu\chi = \chi \otimes \chi$ and define $p \in \mathrm{Der}_K G$ by the formula

$$p = (\xi - \xi^q)\frac{\partial}{\partial \xi} + \xi\chi\frac{\partial}{\partial \chi}$$

Since $\xi - \xi^q$ and ξ belong to $X_a(G)$ and $\frac{\partial}{\partial \xi}$, $\chi\frac{\partial}{\partial \chi}$ belong to L(G) it follows by (1.9) that $p \in P(G)$. Now formula (*) implies that

$$p^q\chi = p\chi \text{ and } p^q(\chi^{-1}) = p(\chi^{-1})$$

On the other hand clearly $p^i\xi = p\xi$ for all $i \geq 2$. Consequently $p \in P(G,\mathrm{fin})$. On the other hand consider the derivation $\tilde{p} \in \mathrm{Der}_K G$,

$$\tilde{p} = (\xi^q - \xi)\frac{\partial}{\partial \xi}$$

Clearly $\tilde{p}\chi = 0$, $\tilde{p}(\chi^{-1}) = 0$, $\tilde{p}^i\xi = (-1)^{i+1}\tilde{p}\xi \in P(G,\mathrm{fin})$. But $\tilde{p} + p \notin P(G,\mathrm{fin})$ because

$$(p + \tilde{p})^i\chi = \xi^i\chi \quad \text{for all } i$$

(3.10) Here is an example of an affine (non-algebraic) K-group scheme G, char K = 0, such that P(G/K,fin) is not a linear subspace of P(G). Take G = Spec A, $A = K[\xi_i, i \in \mathbf{N}]$, $\mu\xi_i = \xi_i \otimes 1 + 1 \otimes \xi_i$, $S\xi_i = -\xi_i$, $\varepsilon\xi_i = 0$ (so G is an "infinite vector group scheme"). Define:

$$p = \sum_{i\geq 1} (\xi_{2i+1}\frac{\partial}{\partial \xi_{2i}} + \xi_{2i}\frac{\partial}{\partial \xi_{2i+1}})$$

$$\tilde{p} = \sum_{i\geq 1} (\xi_{2i}\frac{\partial}{\partial \xi_{2i-1}} + \xi_{2i-1}\frac{\partial}{\partial \xi_{2i}})$$

It easy to see using (1.8) that $p, \tilde{p} \in P(G/K,\mathrm{fin})$; but $p + \tilde{p} \notin P(G/K,\mathrm{fin})$ since for any $i \geq 1$, $(p + \tilde{p})^1(\xi_1)$ is a linear form in which ξ_{i+1} occurs with coefficient 1.

EXAMPLES

It is appropriate to examine here the Examples 1,2,3 given after (1.9) from the viewpoint of local finiteness.

Consider first $G = G_a^n$ viewed as an algebraic D-groups ($D = K[P]$, $P = Kp$) via the derivation $\tilde{p} = p^* + \Sigma a_{ij}\xi_i\frac{\partial}{\partial \xi_j}$ as in (1.26). Then G is locally finite because for each $d \geq 1$ the K-space of all polynomials in $K[\xi_1, \ldots \xi_n]$ of degree $\leq d$ is $\tilde{p}$ - stable.

Consider now $G = G_m \times G_a$ viewed as an algebraic D-group via the derivation $\tilde{p} = p^* + \xi\chi\frac{\partial}{\partial \chi}$. Then G is not locally finite because the D-submodule of $K[\chi,\chi^{-1}\xi]$ generated by χ is infinite dimensional. Finally, consider $G = A \times G_a$ viewed as an algebraic D-group via the derivation $\tilde{p} = p^* + \xi y\frac{\partial}{\partial x}$. Then G is not locally finite; for if it was then by (3.7) it would split over a D-field extension of K, this forcing the subfield K(A)=K(x,y) of $K(A \times G_a) = K(x,y\xi)$ to be preserved by $\tilde{p}$. This is however not the case since $\tilde{p}x = \xi y \notin K(A)$.

(3.11) A D-group scheme G will be called split if there is a D-group scheme isomorphism $G \simeq K \otimes_{K^D} G_0$ where G_0 is a K^D-group scheme and $K \otimes_{K^D} G_0$ is viewed as a D-group scheme with P acting on $K \otimes 1$ via ∂ and acting trivially on $\mathcal{O}_{G_0}$. By (3.6), 1) a D-group scheme G is split if and only if it is so as a D-scheme. So Proposition (3.7) holds if we replace "D-variety" by "irreducible algebraic D-group".

For the reader familiar with $[K_1]$ let's also make the remark that if P has a finite commuting basis $p_1, \ldots, p_m$ and if G is a split irreducible algebraic D-group ($G \simeq K \otimes_{K^D} G_0$) and V is a D-principal homogenous space for G then upon viewing the fields K and $K < V >$ as Δ - fields with derivations $\partial(p_1), \ldots, \partial(p_m)$ we have that the extension $K < V >/K$ is G_0 - primitive in Kolchin's sense $[K_1]$ (cf. also $[B_1]$ p. 26).

We close by a general remark on the link between P(G/K,fin) and the map λ_G in (1.23) (for an extension of this result confer with (IV.2) below).

(3.12) PROPOSITION. Assume char K = 0 and G is an affine algebraic K-group. Then the map

$$\lambda : L(\text{Aut}\, G) \to (\underline{\text{Aut}}\, G) = P(G/K)$$

is injective and its image equals P(G/K,fin).

Proof. Start with a preparation. Assume W is a finite dimensional vector subspace of (G) generating (G) as a K-algebra and let

$$\underline{\text{Aut}}(G,W) : \{\text{affine K-schemes}\} \to \{\text{groups}\}$$

be the subfunctor of $\underline{\text{Aut}}\, G$ defined by

$$S = \text{Spec}\, B \to \{f \in \text{Aut}\,(G \times S/S);\ f^* : \mathcal{O}(G) \otimes B \to \mathcal{O}(G) \otimes B \text{ preserves } W \otimes B\}$$

We claim that $\underline{\text{Aut}}(G,W)$ is representable; note that the affine group scheme Aut(G,W) representing it will be reduced by [Sw] p. 280. To prove our claim let W_1 be the intersection of all sub-K-coalgebras of (G) containing $W_0 = W + SW$; by [Sw] W_1 is a finite dimensional coalgebra. Now define inductively the increasing sequence of subspaces W_i of (G) be the formula $W_{i+1} = \mu(W_i \otimes W_i)$ for $i \geq 1$ and define functors $\underline{A}_0, \underline{A}_1, \underline{A}_2, \ldots$

$$\underline{A}_i : \{\text{affine K-schemes}\} \to \{\text{groups}\}$$

as follows. We let $\underline{A}_0(\text{Spec}\, B)$ be the group of all B-linear automorphisms σ_0 of $W_0 \otimes B$ such that $\sigma_0 S_B = S_B \sigma_0$ where $S_B = S \otimes 1_B$. For $i \geq 1$, let $\underline{A}_i(\text{Spec}\, B)$ be the group of all B-linear automorphisms σ_i of $W_i \otimes B$ such that

$$\sigma_i(W_{i-1} \otimes B) = W_{i-1} \otimes B$$

$$\sigma_i \,\Big|\, W_{i-1} \otimes B \in \underline{A}_{i-1}(\operatorname{Spec} B) \quad \text{and}$$

$$\sigma_i \mu_B = \mu_B(\sigma_{i-1} \otimes \sigma_{i-1}) \quad \text{where} \quad \mu_B = \mu \otimes 1_B$$

We have canonical restriction maps $\underline{A}_i \to \underline{A}_{i-1}$ for $i \geq 1$. Now clearly all $\underline{A}_i$'s are representable by affine algebraic K-groups A_i hence we have a projective system

$$\ldots \to A_i \to A_{i-1} \to \ldots \to A_1 \to A_0$$

One checks that $\underline{\mathrm{Aut}}(G,W) = \varprojlim \underline{A}_i$. Consequently $\underline{\mathrm{Aut}}(G,W)$ is represented by $\operatorname{Spec}(\varinjlim (A_i))$ and our claim is proved. Let's prove that $\operatorname{Im}\lambda = P(G/K,\mathrm{fin})$. The inclusion "$\subset$" is clear. Conversely, if $p \in P(G/K,\mathrm{fin})$ we may choose W above such that $pW \subset W$. Then $\mathrm{id} + \varepsilon p \in \underline{\mathrm{Aut}}(G,W)\,(\operatorname{Spec} K[\varepsilon])$, hence we get a morphism $f : \operatorname{Spec} K[\varepsilon] \to \mathrm{Aut}(G,W)$ such that $\mathrm{id} + \varepsilon p = f^*\phi_{G,W}$ where $\phi_{G,W}$ is the universal $\mathrm{Aut}(G,W)$-automorphism of $G \times \mathrm{Aut}(G,W)$. Now since $\mathrm{Aut}(G,W)$ is reduced and since $\underline{\mathrm{Aut}}\,G$ restricted to {reduced K-schemes} is representable by a locally algebraic K-group [BS] $\mathrm{Aut}\,G$ there exists a morphism $h : \mathrm{Aut}(G,W) \to \mathrm{Aut}\,G$ such that $\phi_{G,W} = h^*\phi_G$ where ϕ_G is the universal $\mathrm{Aut}\,G$-automorphism of $G \times \mathrm{Aut}\,G$. Consequently $\mathrm{id} + \varepsilon p = (hf)^*\phi_G$ hence $p \in \operatorname{Im}\lambda$.

Finally, let's prove that λ is injective. Let $\mathrm{Aut}^oG = \operatorname{Spec} R$ (by [BS] Aut^oG is affine !); we may choose a finite dimensional subspace W of $\mathcal{O}(G)$ generating $\mathcal{O}(G)$ as a K-algebra such that $\phi_G^*(W \otimes R) = W \otimes R$ (where $\phi_G^* : \mathcal{O}(G) \otimes R \to \mathcal{O}(G) \otimes R$ is induced by ϕ_G). Exactly as above there exists a morphism $h : \mathrm{Aut}(G,W) \to \mathrm{Aut}\,G$ such that $h^*\phi_G = \phi_{G,W}$. There is also a natural morphism $c : \mathrm{Aut}^oG \to \mathrm{Aut}(G,W)$ defined at the level of S-points by $(\mathrm{Aut}^oG)(S) \to \mathrm{Aut}(G,W)(S)$, $s \mapsto \tilde{s}$ where $s^*\phi_G = \tilde{s}^*\phi_{G,W}$. Consider the affine group scheme $A = \mathrm{Aut}(G,W) \times_{\mathrm{Aut}\,G} \mathrm{Aut}^oG$; note that the projection $p_1 : A \to \mathrm{Aut}(G,W)$ is a closed embedding. Now the map

$$A \xrightarrow{p_2} \mathrm{Aut}^oG \xrightarrow{c} \mathrm{Aut}(G,W)$$

equals the map

$$A \xrightarrow{p_1} \mathrm{Aut}(G,W)$$

for if $(f,f') \in A(S)$ is an S-point of A we have $hf = if'$ (where $i : \mathrm{Aut}^oG \to \mathrm{Aut}\,G$ is the inclusion) so the image of (f,f') via $(cp_2)(S)$ is a map $s \in \mathrm{Aut}(G,W)(S)$ such that $s^*\phi_{G,W} = f'^*i^*\phi_G = f^*h^*\phi_G = f^*\phi_{G,W}$; consequently $s = f$ by universality of $\mathrm{Aut}(G,W)$. We get that p_2 is a closed embedding so A is an algebraic affine group. Since the map $p_2 : A \to \mathrm{Aut}^oG$ induces a bijection at the level of K-points this map is an isomorphism. Consequently c is a closed embedding hence Aut^oG is a subfunctor of $\mathrm{Aut}(G,W)$ hence of $\underline{\mathrm{Aut}}\,G$ and injectivity of λ follows.

CHAPTER 2. AFFINE D-GROUP SCHEMES

Throughout this chapter K is algebraically closed, P is any Lie K/k-algebra and D = K[P].

1. The analytic method

Our first main result is:

(1.1) THEOREM. Assume char K = 0 and let G be an irreducible affine algebraic K-group. Then $K^{P(G)}$ is a field of definition for G (hence it coincides with K_G).

(1.2) The above statement fails for non-linear G, cf. Chapter 3. Our proof of (1.1) will be analogue to that of Theorem (1.1) in Chapter 2 of $[B_1]$ in the sense that we are going to use "birational quotients", a "Kodaira-Spencer map" and an analytic ingredient. In $[B_1]$ the analytic ingredient was the versal deformation of a compact complex space. Here the analytic ingredient is a combination of Theorems (1.3) and (1.4) below (due to Hamm and Hochschild--Mostow respectively). To state the first theorem let's fix some notations. Assume $\Pi : \mathcal{G} \to \mathcal{X}$ is an analytic family of connected complex Lie groups, i.e. a map of analytic $\mathbf{C}$-manifolds, having connected fibres, such that one is given analytic $\mathcal{X}$- maps

$$\mu : \mathcal{G} \times_{\mathcal{X}} \mathcal{G} \to \mathcal{G}$$

$$S : \mathcal{G} \to \mathcal{G}$$

$$\varepsilon : \mathcal{X} \to \mathcal{G}$$

satisfying the usual axioms of multiplication, inverse and unit. Assume moreover that a) $\mathcal{X}$ is simply connected and $v_1, \dots, v_m$ are commuting vector fields on $\mathcal{X}$ giving at each point a basis of the tangent space and b) $v_1, \dots, v_m$ can be lifted to commuting vector fields $w_1, \dots, w_m$ on $\mathcal{G}$ such that μ, S, ε agree with $w_1, \dots, w_m$ in the sense that for each $w = w_i$ we have:

1) $(T_{(g_1,g_2)}\mu)(w(g_1), w(g_2)) = w(\mu(g_1,g_2))$ for any $(g_1,g_2) \in \mathcal{G} \times_{\mathcal{X}} \mathcal{G}$

2) $(T_g S)(w(g)) = w(Sg)$ for any $g \in \mathcal{G}$

3) $(T_x \varepsilon)(v(x)) = w(\varepsilon(x))$ for any $x \in \mathcal{X}$

Then we have

(1.3) THEOREM. (Hamm, [Ha]). Under the assumptions above there exists an analytic $\mathcal{X}$-isomorphism $\phi : \mathcal{G}_0 \times \mathcal{X} \to \mathcal{G}$ (where $\mathcal{G}_0$ is some fibre of Π) which above each point of $\mathcal{X}$ is a group isomorphism and such that upon letting v_i^* be the trivial lifting of v_i from $\mathcal{X}$ to $\mathcal{G}_0 \times \mathcal{X}$ we have $(T\phi)(v_i^*) = w_i$ for all i.

(1.4) THEOREM. (Hochschild-Mostow [HM_1]). Let G_1, G_2 be two irreducible affine algebraic $\mathbb{C}$-groups. If G_1^{an} and G_2^{an} are isomorphic (as analytic Lie groups) then G_1 and G_2 are isomorphic (as algebraic groups).

(1.5) REMARK. The above statement fails for non-linear groups (cf. [HM_1][Se]).

Theorem (1.4) is a consequence of the theory developed in [HM_1] and no idea of proof will be indicated here. We shall however include the proof (due to Hamm) of (1.3) since it is completely elementary and fairly elegant:

(1.6) Proof of (1.3) (Hamm). By Frobenius, for any $x_0 \in \mathcal{X}$ and any $g_0 \in \Pi^{-1}(x_0) = \mathcal{G}_0$ there exists a neighbourhood $\mathcal{V}_{g_0}$ of x_0 in $\mathcal{X}$, a neighbourhood $\mathcal{W}_{g_0}$ of g_0 in $\mathcal{G}_0$ and an analytic map $\psi : \mathcal{W}_{g_0} \times \mathcal{V}_{g_0} \to \mathcal{G}$ over $\mathcal{X}$ such that $T\phi$ takes v_i^* (= trivial lifting of v_i from $\mathcal{X}$ to $\mathcal{W}_{g_0} \times \mathcal{V}_{g_0}$) into w_i. A triple $(\mathcal{V}_{g_0}, \mathcal{W}_{g_0}, \psi)$ as above will be called a "local solution" at g_0. It is sufficient to show that for a given x_0, the various $\mathcal{V}_{g_0}$'s appearing in the local solutions at various points $g_0 \in \mathcal{G}_0$ can be chosen to contain a fixed open neighbourhood of x_0. Let $e_0 = \varepsilon(x_0)$ and consider the set Σ of all $g \in \mathcal{G}_0$ such that there exists a local solution $(\mathcal{V}_g, \mathcal{W}_g, \psi)$ at g with $\mathcal{V}_{e_0} \subset \mathcal{V}_g$. One easily checks that Σ is an open subgroup of $\mathcal{G}_0$ (local solutions can be "multiplied" and "inverted" using μ and S) hence $\Sigma = \mathcal{G}_0$ by connectivity, which proves the theorem.

Next we need some facts about isomorphism of Lie algebras. First "recall" the following trivial representability result:

(1.7) LEMMA. Let R be a ring and L, L' two Lie R-algebras which are free and finitely generated as R-modules. Then the functor $\underline{Iso}_{L,L'}$, from {R-algebras} to {sets} defined by

$$\underline{Iso}_{L,L'}(\tilde{R}) = \{\tilde{R} \text{ - Lie algebra isomorphisms } \tilde{R} \otimes_R L \simeq \tilde{R} \otimes_R L'\}$$

is representable by a finitely generated R-algebra (call it $Iso_{L,L'}$).

Exactly as in [B_1] pp. 35-36 the above Lemma implies the following:

(1.8) LEMMA. Let C be an algebraically closed field, S an affine C-variety and L a Lie $\mathcal{O}(S)$-algebra which is free and finitely generated as an $\mathcal{O}(S)$-module. Then there is a constructible subset $Z \subset S \times S$ such that for any $s_1, s_2 \in S(C)$ we have $(s_1, s_2) \in Z(C)$ if and only if the Lie C-algebras $L \otimes_{\mathcal{O}(S)} C(s_1)$ and $L \otimes_{\mathcal{O}(S)} C(s_2)$ are isomorphic.

Next we have:

(1.9) LEMMA. Assume C, S, L are as in (1.8), let K be an algebraically closed extension of C(S) and assume K_L (= smallest algebraically closed field of definition of $L \otimes_{\mathcal{O}(S)} K$ between C and K, cf. (I. 1.27)) equals the algebraic closure of C(S) in K. Then there exists an open subset

$S_0 \subset S$ such that for any $s_0 \in S_0(C)$ the set $\{s \in S(C);\ L \otimes C(s) \simeq L \otimes C(s_0)\}$ is finite.

Proof. By (1.8) and $[B_1]$, (1.13), p. 36 there exists an affine open set $S_1 \subset S$ and a dominant morphism of affine C-varieties $\psi : S_1 \to M$ such that for any $s_1 \in S_1(C)$ we have

$$\psi^{-1}\psi(s_1) = \{s \in S_1(C);\ L \otimes C(s) \simeq L \otimes C(s_1)\}$$

if $\dim M = \dim S_1$ we are done. Assume $\dim M < \dim S_1$. Then we use an argument similar to [V] p. 576. Choose a closed subvariety $N \subset S_1$ with $\dim N = \dim M$, let L_N be the pull-back of L on N, let L' be the pull-back of L_N on the affine scheme $\tilde{S}_1 = S_1 \times_M N$ and let L" be the pull-back of L on $\tilde{S}_1$. Then for any C-point $x \in \tilde{S}_1(C)$ one checks that $L' \otimes C(x) = L'' \otimes C(x)$. By representability of $\underline{\mathrm{Iso}}_{L',L''}$ (1.7) there is a generically finite dominant morphism of finite type of affine schems $Y \to \tilde{S}_1$ with Y integral such that the pull-backs of L' and L" on Y are Y-isomorphic. Since $Y \to S_1$ is generically finite, one can embed C(Y) over $C(S_1) = C(S)$ into K and we get that C(N) is a field of definition for $L \otimes K$ between C and K, contradicting our hypothesis. The lemma is proved.

Next we need a Kodaira-Spencer map for irreducible affine algebraic groups G over non-necessarily algebraically closed fields F of characteristic zero containing k. Consider the following complex (where $A = \mathcal{O}(G)$):

$$0 \to \mathrm{Der}_F(A,A) \xrightarrow{\partial_1} \mathrm{Der}_F(A, A \otimes_F A) \xrightarrow{\partial_2} \mathrm{Der}_F(A, A \otimes_F A \otimes_F A)$$

where

$$\partial_1(p) = \mu p - (p \otimes 1 + 1 \otimes p)\mu, \qquad p \in \mathrm{Der}_F(A,A)$$

$$\partial_2(d) = (d \otimes 1)\mu - (1 \otimes d)\mu + (\mu \otimes 1)d - (1 \otimes \mu)d, \qquad d \in \mathrm{Der}_F(A, A \otimes_F A)$$

and define $H^i(G/F)$ for $i = 1,2$ as the i-th homology space of this complex (one can identify $H^i(G/F)$ with the corresponding Hochschild cohomology spaces of the adjoint representation of G, cf. [DG] p. 192, but we won't need this fact). Clearly $H^i(G \otimes_F F_1/F_1) \simeq H^i(G/F) \otimes_F F_1$ for any field extension F_1/F. Moreover we can define P(G) and P(G/F) exactly as in (I. 1.1) in case our ground field is not algebraically closed.

(1.10) LEMMA. There is an exact sequence

$$0 \to P(G/F) \to P(G) \xrightarrow{\partial} \mathrm{Der}_k F \xrightarrow{\rho} H^2(G/F)$$

where ρ is compatible with field extensions F_1/F. Moreover, we have an identification

$$P(G/F) \simeq H^1(G/F)^S$$

(where $H^1(G/F)^S$ is the fixed part of $H^1(G/F)$ under the involution $v \mapsto SvS$ of $\mathrm{Der}_F(A,A)$ which is easily seen to preserve $H^1(G/F)$).

Proof. The assertion about $H^1(G/F)^S$ is clear. Let's define ρ. Since G/F is smooth and affine, any derivation $p \in \mathrm{Der}_k F$ can be lifted to a k-derivation $\tilde{p}$ of $A = \mathcal{O}(G)$.

Then one checks immediately that

$$\mu\tilde{p} - (\tilde{p} \otimes 1 + 1 \otimes \tilde{p})\mu \in \operatorname{Ker}(\partial_2) \subset \operatorname{Der}_F(A, A \otimes_F A)$$

and the class of this derivation in $H^2(G/F)$ does not depend on the choice of the lifting $\tilde{p}$; call this class $\rho(p)$. We must check that

$$\operatorname{Ker} \rho = \operatorname{im}(P(G) \to \operatorname{Der}_k F)$$

The inclusion "$\subset$" is clear. Conversely, if $\rho(p) = 0$ then there is a lifting $\tilde{p}$ of p, $\tilde{p} \in \operatorname{Der}_k A$ such that $\mu\tilde{p} = (\tilde{p} \otimes 1 + 1 \otimes \tilde{p})\mu$. Then one immediately checks that

$$\mu(S\tilde{p}S) = [(S\tilde{p}S) \otimes 1 + 1 \otimes (S\tilde{p}S)]\mu$$

Putting $\hat{p} = (1/2)(\tilde{p} + S\tilde{p}S)$ we see that $\hat{p}$ lifts p and belongs to P(G) so our lemma is proved.

(1.11) COROLLARY. If F_1/F is an exension of algebraically closed fields and G is an irreducible affine algebraic F-group then:

$$P(G \otimes_F F_1/F_1) \simeq P(G/F) \otimes_F F_1 \qquad \text{and}$$

$$P(G \otimes_F F_1/F_1, \text{fin}) \simeq P(G/F, \text{fin}) \otimes_F F_1 .$$

Proof. The first isomorphism follows from (1.10). The second follows from (1.3.12).

(1.12) THEOREM. Assume char K = 0 and G is an irreducible affine algebraic K-group. Let C be an intermediate field between k and K and let K_0 be the smallest algebraically closed field of definition for G between C and K ($G \simeq G_0 \otimes_{K_0} K$ for some K_0-group G_0). Then the map $\rho_0 : \operatorname{Der}_C K_0 \to H^2(G_0/K_0)$ is injective.

REMARK. The field C in the above statement has only an auxiliary role in the proof. The interesting case of course is that when C = k (then $K_0 = K_G$. cf. (I. 1.27)).

We shall give first the proof of (1.12) when C is the field $\mathbf{C}$ of complex numbers. By (I. 1.27) K_0 equals the smallest algebraically closed field of definition for L(U) (U the unipotent radical of G) between C and K. There exist group schemes $\tilde{G} \to S$ and $\tilde{U} \to S$ (S an affine $\mathbf{C}$-variety, $\tilde{U}$ a closed subgroup scheme of $\tilde{G}$) such that K_0 is the algebraic closure of $K_1 = C(S)$, $\tilde{G} \otimes_S K_0 = G_0$, $\tilde{U} \otimes_S K_0 = U_0$ (U_0 unipotent radical of G_0) and the fibres of $\tilde{U} \to S$ are the unipotent radicals of the fibres of $\tilde{G} \to S$. We may assume the relative Lie algebra $L(\tilde{U}/S)$ is a free $\mathcal{O}(S)$-module. Assume ρ_0 is not injective. Then $\rho_1 : \operatorname{Der}_C K_1 \to H^2(G_1/K_1)$ is not injective (where $G_1 = \tilde{G} \otimes_S K_1$). By (1.10) there exists $p \in \operatorname{Der}_C K_1$, $p \neq 0$ which lifts to a derivation $\tilde{p} \in P(G_1)$. Now both p and $\tilde{p}$ can be viewed as rational vector fields on the $\mathbf{C}$-varieties S and $\tilde{G}$ respectively. Shrinking S we may assume both p and $\tilde{p}$ are regular everywhere. Now by (1.9) there exists a Zarisky open set $S_0 \subset S$ such that for any $s_0 \in S_0(\mathbf{C})$ the set

$$\Sigma_{s_0} = \{ s \in S_0(\mathbf{C});\ L(U_s) \simeq L(U_{s_0}) \}$$

is finite where $U_s = \tilde{U} \times_S \mathbf{C}(s)$. Let $\mathcal{X}$ be an analytic disk in S_0^{an} which is an integral submanifold for p and let $\mathcal{G} = \tilde{G}^{an} \times_{S^{an}} \mathcal{X}$. Then $\mathcal{G}$ is an integral submanifold of $\tilde{G}^{an}$ for $\tilde{p}$. By (1.3) all fibers of $\mathcal{G} \to \mathcal{X}$ are pairwise isomorphic as complex Lie Groups. By (1.4) the fibres of $\tilde{G} \to S$ above the points in $\mathcal{X}$ are pairwise isomorphic as algebraic groups; in particular all Lie algebras $L(U_s)$ with $s \in \mathcal{X}$ are pairwise isomorphic, this contradicting the finitness of Σ_{s_0} for $s_0 \in S(\mathbf{C})$. The theorem (1.12) is proved for $C = \mathbf{C}$.

(1.13) Proof of Theorem (1.1). It is sufficient to prove that for any Lie K/k-subalgebra P of P(G), K^P is a field of definition of G.

Case 1. K^P is uncountable. Then we can assume $\mathbf{C} \subset K^P$. Let K_0 be the smallest algebraically closed field of definition for G between $\mathbf{C}$ and K, $G = G_0 \otimes_{K_0} K$. We may conclude by inspecting the diagram with exact rows and colomns (cf. (1.12), case $C = \mathbf{C}$ and (1.10))

$$\begin{array}{ccccc}
 & & 0 & & \\
 & & \downarrow & & \\
 & & \mathrm{Der}_{K_o} K & & 0 \\
 & & \psi \downarrow & & \downarrow \\
P(G) & \xrightarrow{\phi} & \mathrm{Der}_{\mathbf{C}} K & \xrightarrow{\rho} & H^2(G/K) \\
 & & \downarrow & & \downarrow \\
0 \longrightarrow & & (\mathrm{Der}_{\mathbf{C}} K_o) \otimes_{K_o} K & \xrightarrow[\rho_o \otimes K]{} & H^2(G_o/K_o) \otimes_{K_o} K \\
 & & \downarrow & & \downarrow \\
 & & 0 & & 0
\end{array}$$

that $\mathrm{Im}\,\phi = \mathrm{Im}\,\psi$ hence that $K_0 \subset K^P$, i.e. that G is defined over K^P.

Case 2. K^P is countable. Then take an embedding $K^P \subset \mathbf{C}$ and conclude exactly as in $[B_1]$ p. 43 (instead of $[B_1]$, Lemma (1.19) use the fact which we already know from $[B_4]$ that the set of algebraically closed fields of definition for a linear algebraic group has a minimum element, cf. also (I. 1.27)).

(1.14) REMARK. Using Theorem (1.1) one can immediately prove Theorem (1.12) for arbitrary C.

Now (1.1) and (I. 1.25) imply:

(1.15) COROLLARY. Assume $\mathrm{char}\, K = 0$ and let G be an irreducible affine algebraic K-group. Then we have an exact split sequence of Lie K/k-algebras:

$$0 \to P(G/K) \to P(G) \xrightarrow{\partial} \mathrm{Der}(K/K_G) \to 0$$

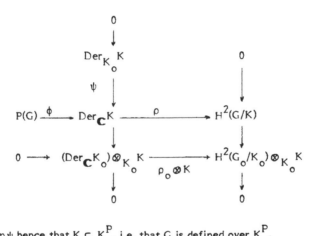

EXAMPLE

Assume P is a Lie K/k-algebra, D = K[P] and $K^D = k \neq K$. Then the "sufficiently general" unipotent algebraic K-group G of sufficiently big dimension doesn't admit any structure of algebraic D-group. Indeed it is known that such a G is not defined over k.

We shall need in what follows a variation on Hamm's result (1.3):

(1.16) LEMMA. Let $\mathcal{G}$ be a connected Lie group and v an analytic vector field on $\mathcal{G}$ such that the multiplication $\mathcal{G} \times \mathcal{G} \to \mathcal{G}$ and the inverse $\mathcal{G} \to \mathcal{G}$ are equivariant (with respect to the vector field (v,v) on $\mathcal{G} \times \mathcal{G}$ and v on $\mathcal{G}$). Then there is a 1-parameter group of analytic group automorphisms $\mathbf{C} \times \mathcal{G} \to \mathcal{G}$ whose derivative at the origin is v.

Proof. Use Hamm's open subgroup argument (1.3) once again to show that there is a disk $0 \in B \subset \mathbf{C}$ such that for all $g \in \mathcal{G}$ there exists an analytic map $\psi_g : B \to \mathcal{G}$, $\psi_g(0) = g$ whose tangent map $T\psi : TB \to T\mathcal{G}$ takes $\frac{d}{dz}$ (z a coordinate in $\mathbf{C}$) into v; this immediately implies the lemma.

Now recall from $[HM_2]$ that the connected component of the group $\mathrm{Aut}(G^{an})$ of analytic group automorphisms of G^{an} has a natural structure of affine algebraic group of which $\mathrm{Aut}^0 G$ is an algebraic subgroup (both being viewed as embedded into GL(L(G))).

(1.17) COROLLARY. Let G be an irreducible affine algebraic $\mathbf{C}$-group. Then $P(G/\mathbf{C}) \subset L(\mathrm{Aut}(G^{an}))$ (as subspaces of $H^0(T_{G^{an}})$).

(1.18) THEOREM. Let char K = 0 and G be an irreducible affine algebraic K-group. Then a derivation in P(G) belongs to P(G,fin) if and only if it preserves the ideal of the unipotent radical of G. In particular P(G,fin) is a Lie K/k-subalgebra of P(G).

We shall provide later (cf. (3.24) of this Chapter) a purely algebraic proof of this result. Here we shall give a proof of it based on the following analytic result:

(1.19) THEOREM. (Hochschild-Mostow $[HM_2]$). If G is a connected affine algebraic $\mathbf{C}$-group then an analytic group automorphism $\phi \in \mathrm{Aut}(G^{an})$ belongs to Aut G if and only if it preserves the unipotent radical U of G.

(1.20) Proof of (1.18). Start with the "only if" part so assume $p \in P(G,\mathrm{fin})$. Then by (I. 3.11) there exists a K[p]-field extension K_1/K such that $K_1^{K[p]} = K^{K[p]}$ (call this field K_0) and such that $K_1 \otimes_K G$ is a split algebraic $K_1[p]$-group so $K_1 \otimes_K G \simeq K_1 \otimes_{K_0} G_0$ for some algebraic K_0-group G_0. Let U_0 be the unipotent radical of G_0; then $K_1 \otimes_{K_0} U_0$ must correspond via the above isomorphism with $K_1 \otimes_K U$. But p clearly preserves the ideal of $K_1 \otimes_{K_0} U_0$ so it will preserve $K_1 \otimes_K U$ hence U itself (by faithfully flatness of the projection $K_1 \otimes_K G \to G$).

Let's prove the "if part". It is easy to reduce the problem to the case $K = \mathbf{C}$. Let now $p \in P(G)$ preserve U hence also L(U) by (I. 1.16). Then by (1.15) we may write $G = K \otimes_{K_0} G_0$

($K_0 = K_G$, G_0 a K_0-group) so $p = p^* + v$ where p^* acts via ∂ on K and acts trivially on G_0 while $v \in P(G/\mathbb{C})$. Since p^* clearly preserves U so does v. By (1.16) there is a 1-parameter group of automorphisms $\mathbb{C} \times G^{an} \to G^{an}$, $(\phi_t)_{t\in\mathbb{C}}$ whose derivative at $t = 0$ is v. Then for each $t \in \mathbb{C}$, ϕ_t preserves U. By (1.19) $\phi_t \in \mathrm{Aut}\,G$ for each t hence by (I. 3.11) $v \in P(G/\mathbb{C},\mathrm{fin})$. But $\mathcal{O}(G)$ is locally finite both as $\mathrm{Der}(K/K_G)$-module and as $P(G/K,\mathrm{fin}) = L(\mathrm{Aut}\,G)$-module (I. 3.12). Since by (1.11) $P(G/K,\mathrm{fin}) = P(G_0/K_0,\mathrm{fin}) \otimes K$ we have

$$[\mathrm{Der}(K/K_G), P(G/K,\mathrm{fin})] \subset P(G/K,\mathrm{fin})$$

it follows that $\mathcal{O}(G)$ is locally finite also as a $\mathrm{Der}(K/K_G) \oplus P(G/K,\mathrm{fin})$-module in particular as a $K[p]$-module which proves the theorem.

EXAMPLE

Let's re-examine Example 2 given after (1.9) so let $G = G_m \times G_a$. We have seen (1.26) that the elements of P(G) have the form

$$\tilde{p} = p^* + a\xi\chi\frac{\partial}{\partial\chi} + b\xi\frac{\partial}{\partial\xi}$$

where $p \in \mathrm{Der}_k K$, $a,b \in K$. According to Theorem (1.18) the elements of P(G,fin) have the form

$$p^* + b\xi\frac{\partial}{\partial\xi}$$

where $p \in \mathrm{Der}_k K$, $b \in K$. Indeed the ideal of the unipotent radical of G is generated by $\chi - 1$ in $K[\chi,\chi^{-1},\xi]$ and the condition that $\tilde{p}$ preserves it is equivalent to $a = 0$.

(1.21) The following concept will play a role in Chapter 5. Assume G is a K-group scheme and assume P(G,fin) is a K-linear subspace of P(G). An ideal in P(G) will be called a representative ideal if:

a) it is a K-linear subspace of P(G/K)

b) it is a complement of P(G,fin) in P(G)

c) it is Aut G - invariant.

As we shall see in Chapter 3, representative ideals may not exist even if G is algebraic and char K = 0. We shall prove here (using $[HM_2]$ once again):

(1.22) PROPOSITION. Assume char K = 0 and G is an irreducible affine algebraic K-group. Then P(G) contains at least one abelian representative ideal.

Our basic ingredient is:

(1.23) THEOREM (Hochschild-Mostow $[HM_2]$). Let G be an irreducible algebraic $\mathbb{C}$-group. Then $\mathrm{Aut}^o(G^{an})$ is the semidirect product of $\mathrm{Aut}^o G$ by some normal algebraic vector subgroup N of $\mathrm{Aut}^o(G^{an})$.

We will also need the following lemma whose proof is "obvious":

(1.24) LEMMA. Let L be a Lie K-algebra of dimension n, L_1 a Lie subalgebra of dimension n_1 and A a locally algebraic K-group acting on L by Lie algebra automorphisms. Assume K_0 is an algebraically closed subfield of K over which all the above data are defined. Let Y denote the subset of all K-points in the Grassmanian X of $(n - n_1)$-planes of L which correspond to subspaces L' of L enjoying the following properties:

1) L' is an abelian subalgebra of L

2) L' is an ideal in L

3) $L_1 \oplus L' = L$

4) L' is A-invariant.

Then Y is locally closed in X(K) in the obvious K_0-topology of X(K).

(1.25) Proof of (1.22). Put $K_0 = K_G$ And let $G = K \otimes_{K_0} G_0$, cf. (1.1). It is sufficient to find an abelian ideal J_0 of the Lie K_0-algebra $P(G_0/K_0)$ complementary to $P(G_0/K_0,\text{fin})$ and Aut G_0-invariant; because then $P(G,\text{fin}) = \text{Der}(K/K_0) \oplus P(G/K,\text{fin})$ (cf. (1.15), (1.18)) hence $J = J_0 \otimes_{K_0} K$ will be an abelian representative ideal in P(G). Now by (1.24) it is sufficient to find and ideal J of P(G/K) which is abelian, complementary to P(G/K,fin) and Aut G-invariant. By (1.24) again we may assume (after replacing K by a field extension of K or by a suitable subfield of K) that K = **C**. But then (1.17) and (1.23) show that viewing P(G/ **C**) as a subalgebra of $L(\text{Aut}(G^{an}))$ we have that $J = P(G/\mathbf{C}) \cap L(N)$ satisfies our requirements (N as in (1.17)). So (1.22) is proved.

EXAMPLE

Assume $G = G_m \times G_a$ as in Example 2 (1.9), (1.26). A representative ideal in P(G) is the K-space J of all derivations

$$a\xi\chi\frac{\partial}{\partial\chi}, \qquad a \in K$$

Indeed Aut $G = G_m$ acts on $K[\chi,\chi^{-1},\xi]$ by leaving χ fixed and multiplying ξ with non-zero scalars in K so J is Aut G-invariant; it is obviously a complement of P(G,fin) in P(G) and it is an ideal in P(G) because

$$[p^*,a\xi\chi\frac{\partial}{\partial\chi}] = (pa)\xi\chi\frac{\partial}{\partial\chi}, \quad p \in \text{Der}_k K, \ a \in K$$

$$[b\xi\frac{\partial}{\partial\chi},a\xi\chi\frac{\partial}{\partial\chi}] = ab\xi\chi\frac{\partial}{\partial\chi}, \ a,b \in K$$

2. Algebraic method: direct products

The aim of this section is to investigate by purely algebraic means the structure of $P(G \times H)$ where G,H are affine (non-necessarily algebraic) K-group schemes (char K arbitrary) with H linearly reductive (i.e. all rational H-modules are completely reducible). Remarkably, with only a minor extra effort we can make our arguments work in the (non-commutative) Hopf algebra context [Sw]. Therefore we shall place ourselves (in this section only!) in this more general frame and freely borrow from [Sw] the Hopf algebra terminology.

As an application we will provide for instance a description of

$$P(G_a^n \times GL_N)$$

in case char K = 0 (recall that in this case GL_N is linearly reductive).

Let A be a Hopf K-algebra (K/k as usual a field extension of arbitrary characteristic with K algebraically closed) and denote by μ, S, ε the comultiplication, antipode and co-unit of A. Define $\Pi(A)$ to be the Lie K/k-algebra of all k-derivations p of A such that $pK \subset K$, $\mu p = (p \otimes 1 + 1 \otimes p)\mu$, $pS = Sp$, $p\varepsilon = \varepsilon p$ (we wrote $\Pi(A)$ and not P(A) to avoid confusion with the space of primitive elements of A [Sw]). Clearly, if G is an affine K-group scheme with Hopf K-algebra $A = \mathcal{O}(G)$ then $\Pi(A) = P(G)$. Our main result will describe $\Pi(A \otimes B)$ where A, B are Hopf K-algebras with B co-semisimple. Note that by a result of Sweedler [Sw] if $B = \mathcal{O}(H)$ for some affine K-group scheme H then B is co-semisimple if and only if H is linearly reductive; if moreover $A = \mathcal{O}(G)$ for some affine K-group scheme G then of course $\Pi(A \otimes B) = P(G \times H)$.

We start with a preparation on centers and cocenters of Hopf algebras.

(2.1) Let C be a K-coalgebra. A coideal I of C is called cocentral if $\overline{\tau}(1 \otimes q)\mu = (q \otimes 1)\mu : C \to (C/I) \otimes C$ where μ is of course the comultiplication, $q : C \to C/I$ is the canonical surjection and $\overline{\tau} : C \otimes (C/I) \to (C/I) \otimes C$ is the twist map (equivalently, if $\mu x - \tau \mu x \in I \otimes C$ for all $x \in C$, where $\tau : C \otimes C \to C \otimes C$ is the twist map). If dim C < ∞, I is cocentral if and only if the subalgebra $(C/I)^0$ of C^0 is central. One easily sees that if dim C < ∞ then there is a minimum concentral coideal I_C of C; indeed let A be the center of C^0 and put $I_C = \mathrm{Ker}(C = C^{00} \to A^0)$. Now we claim that any co-algebra C (possibly of infinite dimension) has a minimum co-central coideal I_C. Indeed, take any family $(C_i)_i$ of subcoalgebras such that $C = \Sigma C_i$ and $\dim C_i < \infty$; then $I_C = \Sigma I_{C_i}$ is easily seen to be the minimum cocentral coideal in C.

(2.2) A coalgebra C will be called cocentral if $I_C = \mathrm{Ker}\, \varepsilon$ (ε = the counit). If dim C < ∞ then C is cocentral if and only if C^0 is a central algebra.

(2.3) Recall from [Sw] p. 161 the following basic properties of simple coalgebras: if C is a coalgebra, $C = \Sigma C_i$, C_i subcoalgebras then any simple subcoalgebra of C lies in one of the C_i's. Recall that a coalgebra is called co-sesimple if it is the sum of its simple subcoalgebras. The above propety implies that if C is co-semisimple then any subcoalgebra of C has a complementary coalgebra and is the sum of its simple subcoalgebras.

(2.4) Let B be a Hopf K-algebra. A Hopf ideal J is called cocentral if it is so as a coideal. Any Hopf algebra B has a minimum cocentral Hopf ideal J_B; indeed put $J_B = B(\Sigma S^n I_B)B$ where I_B is the minimum cocentral coideal in B. The quotient $B^c = B/J_B$ is called the cocenter of B.

(2.5) Assume in (2.4) above that B^c is a group algebra and let $B = \oplus B_\chi$ be, the $X_m(B^c)$-gradation corresponding to the coaction of B^c on B on the right $\rho : B \to B \otimes B^c$ (in analogy with (I. 1.3) we denoted by $X_m(A)$ rather than by G(A) as in [Sw] the group of group-like elements of A); since J_B is cocentral, this gradation coincides with the gradation corresponding to the left coaction $\lambda : B \to B^c \otimes B$. We claim that all B_χ's are subcoalgebras of B. Indeed $\mu : B \to B \otimes B$ is equivariant with respect to the right coactions of B^c on B and $B \otimes B$ (for the latter take the coaction ρ on the second factor); in particular $\mu(B_\chi) \subset B \otimes B_\chi$. Similarily μ is equivariant with respect to the left coaction of B^c on B and $B \otimes B$ (for the latter take the coaction λ on the first factor) and get $\mu(B_\chi) \subset B_\chi \otimes B$; consequently $\mu(B_\chi) \subset B_\chi \otimes B_\chi$ and our claim is proved.

Note also that $B_\chi \neq 0$ for all $\chi \in X_m(B^c)$; indeed, if $b \in B$ lies above $\chi \in X_m(B^c)$ then b_χ, the χ - homogeneous piece of b, also lies above χ, hence $b_\chi \neq 0$.

(2.6) A Hopf algebra is called co-semisimple [Sw] if it is so as a coalgebra. In [Sw] p. 294 several characterizations of co-semisimple Hopf algebras are given; in particular as already noted commutative co-semisimple Hopf algebras correspond precisely to linearly reductive affine group schemes. Note that if B is cosemisimple then the cocenter B^c is a group algebra. Indeed write $B = \oplus C_i$, C_i simple (cocentral) subcoalgebras. Then $B/I_B = \oplus(C_i/I_{C_i}) = \oplus(C_i/\ker \varepsilon_{C_i})$ is generated by group-like elements hence so will be B^c (being a quotient of B/I_B).

(2.7) Let's discuss the notion of center of Hopf algebra. A subset of a Hopf K-algebra A will be called central if each element of it commutes with all elements of A. Then A contains a maximum central sub Hopf algebra ${}^cA$ (we let ${}^cA$ be the maximum element of the family of all central, S-stable subcoalgebras of A). We call ${}^cA$ the center of A; it is contained (but apriori not equal to) the center of the underlying algebra of A. Note that if $X_a(A)$ respectively $X_a({}^cA)$ denote the spaces of primitive elements of A and ${}^cA$ respectively, then $X_a({}^cA)$ consists precisely of the central elements of $X_a(A)$ (once again we used the notation $X_a(A)$ instead of P(A) as in [Sw] in analogy with our notations in (I. 1.3)).

(2.8) Let A and B be two Hopf K-algebras. Then $A \otimes B$ has a natural structure of Hopf algebra. As in (I. 1.19) there is a K-linear projection $\Pi(A \otimes B) \to \Pi(A) \times_{\mathrm{Der}_k K} \Pi(B)$ over $\mathrm{Der}_k K$ defined by $p \to (p_A, p_B)$ where $p_A = (1_A \otimes \varepsilon_B) p i_A$ (where $i_A : A \to A \otimes B$, $i_A(a) = a \otimes 1$) and p_B is defined similarily. Our projection admits a section (which is a Lie K/k-algebra map) defined by $(p_1, p_2) \to p_1 \otimes 1 + 1 \otimes p_2$. We shall usually identify $\Pi(A) \times_{\mathrm{Der}_k K} \Pi(B)$ with its image in $\Pi(A \otimes B)$.

(2.9) Next assume in (2.8) above that the cocenter B^c is a group algebra; by (2.6) this is the case for instance when B is cosemisimple. Then we shall define a remarkable Lie K-subalgebra $\Pi(B : A)$ of $\Pi(A \otimes B)$ lying in the kernel of the projection $\Pi(A \otimes B) \to \Pi(A) \times_{\mathrm{Der}_k K} \Pi(B)$. We proceed as follows: consider the $X_m(B^c)$-gradation $B = \oplus B_\chi$ on B defined by the coaction of B^c on B (cf. (2.5)) and consider also the space $X_a({}^cA)$ of primitive elements of ${}^cA$. Then for any group homomorphism $a \in \mathrm{Hom}(X_m(B^c), X_a({}^cA))$ define the K-linear map $E_a : A \otimes B \to A \otimes B$ by the formula

$$E_a(x\otimes b)=\sum_\chi xa(\chi)\otimes b_\chi,\quad x\in A,\quad b\in B$$

where $b=\Sigma b_\chi$ is the decomposition of $b\in B$ into homogeneous pieces. Then clearly E_a is an A-derivation and using the fact that all B_χ's are subcoalgebras of B one checks the fact that $E_a\in\Pi(A\otimes B)$. We defined a linear map

$$E:\mathrm{Hom}(X_m(B^c),X_a({}^cA))\to\Pi(A\otimes B)$$

whose image will be denoted by $\Pi(B:A)$. Since $B_\chi\neq 0$ for all χ (cf. (2.5)) it follows that $\mathrm{Hom}(X_m(B^c),X_a({}^cA))\simeq\Pi(B:A)$ is an abelian Lie K-subalgebra of $\Pi(A\otimes B)$.

Our main result is the following:

(2.10) THEOREM. Let A and B be Hopf K-algebras with B cosemisimple. Then

$$\Pi(A\otimes B)=(\Pi(A)x_{\mathrm{Der}_kK}\Pi(B))\oplus\Pi(B:A)$$

(2.11) COROLLARY. Let G and H be affine K-group schemes with H linearly reductive. Then

$$P(G\times H)\simeq(P(G)x_{\mathrm{Der}_kK}P(H))\oplus\mathrm{Hom}_{gr}(X_m(Z(H)),X_a(G))$$

(2.12) Proof of (2.10). We must show that every $p\in\mathrm{Ker}(\Pi(A\otimes B)\to\Pi(A)x_{\mathrm{Der}_kK}\Pi(B))$ lies in $\Pi(B:A)$. Clearly p is K-linear.

Step 1. We show that p is an A-derivation. It is sufficient to check that $p(A\otimes K)\subset A\otimes K$. By (2.3) $B=K\oplus M$ for some coalgebra M. Consequently for any $a\in A$ we may write $p(a\otimes 1)\in a_0\otimes 1+A\otimes M$ with $a_0\in A$. We get that

$$\mu(p(a\otimes 1))\in\mu(a_0\otimes 1)+A\otimes M\otimes A\otimes M,\quad \mu(a_0\otimes 1)\in A\otimes K\otimes A\otimes K$$

On the other hand

$$\mu(p(a\otimes 1))-(p\otimes 1+1\otimes p)\mu(a\otimes 1)\subset(p\otimes 1+1\otimes p)(A\otimes K\otimes A\otimes K)\subset$$

$$\subset A\otimes K\otimes A\otimes K+A\otimes M\otimes A\otimes K+A\otimes K\otimes A\otimes M$$

we get that $\mu(p(a\otimes 1))=\mu(a_0\otimes 1)$ hence $p(a\otimes 1)=a_0\otimes 1$ and our claim is proved.

Step 2. We show that for any subcoalgebra C of B we have $p(K\otimes C)\subset A\otimes C$. Indeed by (2.3) C has a complementary coalgebra C'. For each $b\in C$ we have

$$p(1\otimes b)\in x_b+A\otimes C',\qquad x_b\in A\otimes C$$

We get

$$\mu(p(1\otimes b)) \in \mu(x_b) + A\otimes C'\otimes A\otimes C', \quad \mu(x_b)\in A\otimes C\otimes A\otimes C$$

On the other hand

$$\mu(p(1\otimes b)) = (p\otimes 1 + 1\otimes p)\mu(1\otimes b) \in (p\otimes 1 + 1\otimes p)(K\otimes C + C\otimes K) \subset$$
$$\subset A\otimes B\otimes K\otimes C + K\otimes C\otimes A\otimes B$$

We get that $\mu(p(1\otimes b)) = \mu(x_b)$ hence $p(1\otimes b) = x_b$ and we are done.

Step 3. We show that for any simple subcoalgebra C of B there exists $a_C \in X_a(^CA)$ such that for all $b \in C$ we have $p(1\otimes b) = a_C\otimes b$.

Start with the following general remark: if C is any simple K-coalgebra and $L : C \to C$ is a K-linear map such that $\mu L = (L\otimes 1_C)\mu = (1_C\otimes L)\mu : C \to C\otimes C$ then L is a scalar multiple of the identity. Indeed, choose $t \in K$ such that $L - t1_C$ is not invertible. The equalities

$$\mu(L - t1_C) = ((L - t1_C)\otimes 1_C)\mu = (1_C\otimes(L - t1_C))\mu$$

show that the image V of $L-t1_C$ is a subcoalgebra of C; since V is not the whole of C it must be zero hence $L = t1_C$ and our remark is proved.

Now let $a^* \in A^o$ and

$$L_{a^*} = (a^*\otimes 1_C)pi_C : C \to A\otimes C \to A\otimes C \to C$$

where $i_C : C \to A\otimes C$, $i_C(x) = 1\otimes x$. We claim that L_{a^*} is a scalar multiple of 1_C. To see this note first that $a^* = (a^*\otimes\varepsilon_A)\mu : A \to K$ which implies that

$$\mu_C(a^*\otimes 1_C) = (a^*\otimes 1_C\otimes\varepsilon_A\otimes 1_C)\mu_{A\otimes C} : A\otimes C \to C\otimes C$$

Now if $x \in C$ we get

$$\mu_C(L_{a^*}x) = \mu_C((a^*\otimes 1)p(1\otimes x)) = (a^*\otimes 1_C\otimes\varepsilon_A\otimes 1_C)\mu_{A\otimes C}p(1\otimes x) =$$
$$= (a^*\otimes 1_C\otimes\varepsilon_A\otimes 1_C)(p\otimes 1_{A\otimes C} + 1_{A\otimes C}\otimes p)\mu_{A\otimes C}(1\otimes x) =$$
$$= (((a^*\otimes 1_C)p)\otimes\varepsilon_A\otimes 1_C)\mu_{A\otimes C}(1\otimes x) + (a^*\otimes 1_C)\otimes((\varepsilon_A\otimes 1_C)p)\mu_{A\otimes C}(1\otimes x)$$

The second term of the latter sum vanishes because p projects into $0 \in \Pi(A)x_{Der_k K}\Pi(B)$. Using the usual sigma notation $\mu_C x = \Sigma x_{(1)}\otimes x_{(2)}$ we get

$$\mu_C(L_{a^*}x) = \Sigma(((a^*\otimes 1_C)p)\otimes\varepsilon_A\otimes 1_C)(1\otimes x_{(1)}\otimes 1\otimes x_{(2)}) =$$
$$= \Sigma(L_{a^*}x_{(1)})\otimes x_{(2)} = (L_{a^*}\otimes 1_C)(\mu_C x)$$

In other words $\mu_C L_{a^*} = (L_{a^*}\otimes 1_C)\mu$. Similarily starting with the equality $a^* = (\varepsilon_A\otimes a^*)\mu$ we get $\mu_C L_{a^*} = (1_C\otimes L_{a^*})\mu$. By the remark we made L_{a^*} must be a scalar multiple of the identity and our claim is proved.

Our claim implies that there exists $a \in A$ such that $p(1 \otimes b) = a \otimes b$ for all $b \in C$. All we have to check now is that $a \in X_a({}^cA)$. Writing

$$0 = \mu_{A \otimes B} p(1 \otimes b) - (p \otimes 1_{A \otimes B} + 1_{A \otimes B} \otimes p)\mu_{A \otimes B} b =$$

$$= (1_A \otimes \tau \otimes 1_B)((\mu_A a - a \otimes 1 - 1 \otimes a) \otimes \mu_B b)$$

where $\tau : A \otimes B \to B \otimes A$ is the twist map, we get $a \in X_a(A)$. Writing $xa \otimes b = p(x \otimes b) = $ $= p((1 \otimes b)(x \otimes 1)) = ax \otimes b$ for $x \in A$ we get $a \in X_a({}^cA)$.

Step 4. We show that whenever C and C' are two simple subcoalgebras of B contained in the same B_χ, we have $a_C = a_{C'}$.

Indeed, by (2.1) and (2.4) the ideal $J_B = \mathrm{Ker}(B \to B^c)$ is generated (as a two-sided ideal) by elements of the form $S^n x$ where $n \geq 0$ and x belongs to the union of all simple subcoalgebras of B. This together with Step 3 shows that the ideal $A \otimes J_B$ is stable under p; in particular p induces a derivation $\bar{p} \in \Pi(A \otimes B^c)$ and we have a commutative diagram

$$\begin{array}{ccccc}
A \otimes B & \xrightarrow{\ \mu\ } & A \otimes B \otimes A \otimes B & \xrightarrow{\ 1 \otimes 1 \otimes 1 \otimes q\ } & A \otimes B \otimes A \otimes B^c \\
\downarrow p & & \downarrow p \otimes 1 + 1 \otimes p & & \downarrow p \otimes 1 + 1 \otimes \bar{p} \\
A \otimes B & \xrightarrow[\ \mu\ ]{} & A \otimes B \otimes A \otimes B & \xrightarrow[\ 1 \otimes 1 \otimes 1 \otimes q\ ]{} & A \otimes B \otimes A \otimes B^c
\end{array}$$

Now clearly the projection of $\bar{p}$ in $\Pi(A) \times_{\mathrm{Der}_k K} \Pi(B^c)$ is still zero, hence by Step 3 applied to $K\chi$ and B^c instead of C and B we get that $\bar{p}(1 \otimes \chi) = a_\chi \otimes \chi$ for some $a_\chi \in A$. We have for any $b \in C$

$$(p \otimes 1 + 1 \otimes \bar{p})(1 \otimes 1 \otimes 1 \otimes q)\mu(1 \otimes b) = (p \otimes 1 + 1 \otimes \bar{p})(1 \otimes b \otimes 1 \otimes \chi) =$$

$$= a_C \otimes b \otimes 1 \otimes \chi + 1 \otimes b \otimes a_\chi \otimes \chi$$

On the other hand

$$(1 \otimes 1 \otimes 1 \otimes q)\mu p(1 \otimes b) = (1 \otimes 1 \otimes 1 \otimes q)\mu(a_C \otimes b) = a_C \otimes b \otimes 1 \otimes \chi + 1 \otimes b \otimes a_C \otimes \chi$$

from which we get $a_C = a_\chi$. Similarily $a_{C'} = a_\chi$ and we are done.

Step 5. Conclusion of the proof. By (2.3) each B_χ is a sum of simple subcoalgebras hence by Step 4 we get a function $X_m(B^c) \to X_a({}^cA)$, $\chi \mapsto a(\chi)$ such that $p(1 \otimes b) = a(\chi) \otimes b$ for all $b \in B_\chi$. To check it is a group homomorphism compute $p(1 \otimes b_1 b_2)$ for $b_1 \in B_{\chi_1}$, $b_2 \in B_{\chi_2}$ with $b_1 b_2 \neq 0$ (note that $B_{\chi_1} B_{\chi_2} \neq 0$ for all χ_1, χ_2 because by (2.5) $0 \neq B_{\chi_2} \subset B_{\chi_1^{-1}} B_{\chi_1} B_{\chi_2}$).

(2.13) REMARK. The proof of Step 2 above shows that for any co-semisimple Hopf K-algebra B, any subcoalgebra C of B and any $p \in \Pi(B)$ we have $pC \subset C$. In particular B is a locally finite D-module where $D = K[\Pi(B)]$.

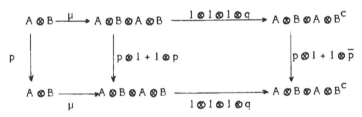

EXAMPLE

The above remark plus (I. 3.2) show that if char K = 0 and G is a reductive linear algebraic K-group then

$$P(G/K) = L(\mathrm{Aut}\, G) = L(\mathrm{Int}\, G) = L(G)/L(Z(G))$$

In particular for instance if $G = GL_N = \mathrm{Spec}\, K[y, \frac{1}{\det y}]$, $y = (y_{ij})$, then P(G/K) has a K-basis consisting of the derivations

$$p_b = \sum_{i,j} \Big(\sum_m b_{im} y_{mj} - \sum_m b_{mj} y_{im}\Big) \frac{\partial}{\partial y_{ij}}$$

where $b = (b_{ij}) \in gl_N(K)$.

Consequently for $G = GL_N$, P(G) consists of all derivations of the form $p^* + p_b$ where $p \in \mathrm{Der}_k K$ and $p^* \in \mathrm{Der}_k K[y, \frac{1}{\det y}]$, lifts p and kills y.

Assume now $G = G_a^n \times GL_N = \mathrm{Spec}\, K[\xi, y, \frac{1}{\det y}]$, $\xi = (\xi_1, \ldots, \xi_n)$. Then by (2.10) P(G) consists of all derivations of the form:

$$\tilde{p} = p^* + p_a + p_b + p_c$$

where $p \in \mathrm{Der}_k K$, p_b is given by the formula above,

$$p_a = \sum_{i,j} a_{ij} \xi_i \frac{\partial}{\partial \xi_j}, \qquad a = (a_{ij}) \in gl_n(K)$$

$$p_c = \Big(\sum_m c_m \xi_m\Big)\Big(\sum_{i,j} y_{ij} \frac{\partial}{\partial y_{ij}}\Big), \qquad c = (c_1, \ldots, c_n),\ c_i \in K$$

It worths noting that P(G,fin) consists of all derivations of the form $p^* + p_a + p_b$. On the other hand the space of derivations of the form p_c is a representative ideal in P(G).

3. Algebraic method: semidirect products

Let M be an affine K-group scheme, T a diagonalizable affine K-group scheme [DG] and $\rho : M \times T \to M$ ($\rho(m,t) = m^t$) a right action of T on M; to simplify discussion we always assume in what follows that $\mathcal{O}(M)$ is an integral domain and $X_m(M) = 1$. Our aim here is to study the structure of $P(M \times_\rho T)$ where $M \times_\rho T$ is the semidirect product of M and T constructed with the help of ρ. This will enable us in particular to re-prove algebraically (1.18), get new information about the map $\ell\nabla : P(G) \to \mathrm{Hom}(X_m(G), X_a(G))$ and compute P(G) in case the radical of G is nilpotent or the unipotent radical of G is commutative. For sake of simplicity we agree from now on the write W(G) instead of $\mathrm{Hom}(X_m(G), X_a(G))$ for any K-group scheme G.

(3.1.) We shall think of $M \times_\rho T$ as having its underlying scheme equal to $M \times T$ while multiplication is given by the formula

$$(m_1,t_1)(m_2,t_2) = (m_1^{t_2}m_2, t_1t_2), \quad m_i \in M(K), \quad t_i \in T(K)$$

hence by the composition

(3.1.1) $$\mu : M \times T \times M \times T \xrightarrow{\tau\times\tau} T \times M \times T \times M \xrightarrow{1\times\tilde{\rho}\times 1} T \times M \times T \times M \longrightarrow$$

$$\xrightarrow{1\times\tau\times 1} T \times T \times M \times M \xrightarrow{\mu_T \times \mu_m} T \times M \xrightarrow{\tau} M \times T$$

where $\tilde{\rho} : M \times T \to M \times T$, $\tilde{\rho}(m,t) = (\rho(m,t),t) = (m^t,t)$ and τ is the twist map. The antipode is given by $(m,t) \to ((m^{-1})^{t^{-1}}, t^{-1})$ hence by the composition

(3.1.2) $$S : M \times T \xrightarrow{S_M \times S_T} M \times T \xrightarrow{\tilde{\rho}} M \times T$$

and the unit is given by

(3.1.3) $$\varepsilon : \operatorname{Spec} K \xrightarrow{\varepsilon_M \times \varepsilon_T} M \times T$$

Let Σ denote the group $X_m(T)$ (hence $\mathcal{O}(T) = K[\Sigma]$ = group algebra on Σ). To $\rho : M \times T \to M$ there corresponds an algebra map still called $\rho : \mathcal{O}(M) \to \mathcal{O}(M) \otimes \mathcal{O}(T)$ hence a Σ-gradation

$$\mathcal{O}(M) = \bigoplus_{\chi\in\Sigma} \mathcal{O}(M)_\chi$$

on $\mathcal{O}(M)$; if we denote by $f_\chi : \mathcal{O}(M) \to \mathcal{O}(M)_\chi \subset \mathcal{O}(M)$ the projection onto the χ-component then $\mu_M f_\chi = \sum_{\chi'\chi''=\chi} (f_{\chi'} \otimes f_{\chi''})\mu_M$ for all χ and we have the formula

(3.1.4) $$\tilde{\rho}(a \otimes \chi) = \Sigma f_{\chi'}(a) \otimes \chi'\chi, \qquad a \in \mathcal{O}(M), \ \chi \in \Sigma$$

where $\tilde{\rho} : \mathcal{O}(M) \otimes \mathcal{O}(T) \to \mathcal{O}(M) \otimes \mathcal{O}(T)$ is induced by $\tilde{\rho}$. we denote by $\Phi(T, (M))$ the set of all weights of T in $\mathcal{O}(M)$ i.e.

$$\Phi(T,\mathcal{O}(M)) = \{\chi \in \Sigma ; \mathcal{O}(M)_\chi \neq 0\} = \{\chi \in \Sigma; f_\chi \neq 0\}$$

Since $\mathcal{O}(M)$ is integral, $\Phi(T,\mathcal{O}(M))$ is a semigroup. For each $\chi \in \Sigma$ put

$$(\operatorname{Der}_k M)_\chi = \{p \in \operatorname{Der}_k M ; p(\mathcal{O}(M))_{\chi'} \subset \mathcal{O}(M)_{\chi'\chi} \text{ for all } \chi' \in \Sigma\} =$$

$$= \{p \in \operatorname{Der}_k M;\ f_{\chi^{-1}\chi'}\, p = p f_{\chi'} \text{ for all } \chi' \in \Sigma\}$$

Of course the sum in $\operatorname{Der}_k M$ of all $(\operatorname{Der}_k M)_\chi$ is direct. We also put $L(M)_\chi = L(M) \cap (\operatorname{Der}_k M)_\chi$ and $P(M)_\chi = P(M) \cap (\operatorname{Der}_k M)_\chi$. Now if $p \in \operatorname{Der}_k(\mathcal{O}(M) \otimes \mathcal{O}(T)) = \operatorname{Der}_k(M \times_\rho T)$ is such that $pK \subset K$ we define for each $\chi \in \Sigma$ endomorphisms $p_\chi \in \operatorname{End}_k \mathcal{O}(M)$ by the formula

$$p_\chi a = (1 \otimes e_\chi) p (a \otimes 1), \quad a \in \mathcal{O}(M)$$

where $e_\chi : \mathcal{O}(T) = \bigoplus_{\chi'} K\chi' \to K\chi \simeq K$ is the natural projection onto the χ-component. In other words the following formula holds:

(3.1.5) $\qquad p(a \otimes 1) = \sum_\chi (p_\chi a) \otimes \chi \quad$ for all $a \in \mathcal{O}(M)$.

One checks easily that $p_1 \in \mathrm{Der}_k \mathcal{O}(M)$ and $p_\chi \in \mathrm{Der}_K \mathcal{O}(M)$ for all $\chi \neq 1$. We put

$$p^0 = \sum_\chi p_\chi \in \mathrm{Der}_k \mathcal{O}(M)$$

Moreover for any χ we put $X_a(M)_\chi = X_a(M) \cap \mathcal{O}(M)_\chi$. It is easy to check that $X_a(M) = \bigoplus X_a(M)_\chi$ and that we have $X_a(M \times_\rho T) = X_a(M)_1 \otimes 1 \subset \mathcal{O}(M) \otimes \mathcal{O}(T)$. Moreover since $X_m(M) = 1$ one checks that $X_m(M \times_\rho T) = 1 \otimes X_m(T) \subset \mathcal{O}(M) \otimes \mathcal{O}(T)$ so we have

$$W(M \times_\rho T) = \mathrm{Hom}(X_m(T), X_a(M)_1)$$

(3.2) LEMMA. Let $p \in \mathrm{Der}_k(M \times_\rho T)$ such that $pK \subset K$ as in (3.1). Then $p \in P(M \times_\rho T)$ if and only if the following conditions are satisfied:

1) For any $\chi \in \Sigma$ we have $p(1 \otimes \chi) = a_\chi \otimes \chi$ for some $a_\chi \in X_a(M)_1$.

2) For any $\chi \in \Sigma$ we have

$$(f_\chi \otimes 1)\mu_M p_1 = (p_1 f_\chi \otimes 1 + \sum_\eta f_{\chi\eta^{-1}} \otimes p_\eta + f_\chi \otimes a_\chi)\mu_M$$

3) For any $\chi, \chi' \in \Sigma$, $\chi \neq 1$ we have

$$(f_{\chi'\chi^{-1}} \otimes 1)\mu_M p_\chi = (p_\chi f_{\chi'} \otimes 1)\mu_M$$

4) $p^0 \epsilon_M = \epsilon_M p^0$

5) For any $\chi \in \Sigma$ we have

$$\sum_\eta S_M p_{\chi^{-1}\eta} S_M f_\chi + a_\chi f_\chi = \sum_\eta f_{\eta\eta^{-1}\chi} p_\eta$$

Proof. 1) is equivalent to the fact that μp and $(p \otimes 1 + 1 \otimes p)\mu$ agree on $1 \otimes \Sigma$ (use(3.1)). We claim that 2)+3) is equivalent to the fact that these maps agree on $\mathcal{O}(M) \otimes 1$. Indeed, for $a \in \mathcal{O}(M)$ we have (using (3.1.1), (3.1.5)):

$$\mu p(a \otimes 1) = (\tau \otimes \tau)(1 \otimes \tilde{\rho} \otimes 1)(1 \otimes \tau \otimes 1)(\mu \otimes \mu)\tau(\sum p_\chi a \otimes \chi) =$$

$$= \sum f_{\chi'}((p_\chi a)_{(1)}) \otimes \chi \otimes (p_\chi a)_{(2)} \otimes \chi\chi'$$

$$(p \otimes 1 + 1 \otimes p)\mu(a \otimes 1) = \sum p_{\chi'}(f_\chi a_{(1)}) \otimes \chi' \otimes a_{(2)} \otimes \chi +$$

$$+ \sum f_\chi a_{(1)} \otimes 1 \otimes p_{\chi'} a_{(2)} \otimes \chi\chi' + \sum f_\chi a_{(1)} \otimes 1 \otimes a_{(2)} a_\chi \otimes \chi$$

Using the fact that $\mathcal{O}(M) \otimes \mathcal{O}(T) \otimes \mathcal{O}(M) \otimes \mathcal{O}(T)$ is a free $\mathcal{O}(M) \otimes \mathcal{O}(M)$-module with basis $1 \otimes \chi \otimes 1 \otimes \chi'$ and identifying coefficients we get our claim.

Condition 4) is equivalent to $p\varepsilon = \varepsilon p$. Finally using formula (3.1.2) and a computation similar to the one above we see that condition 5) is equivalent to $pS = Sp$.

(3.3) COROLLARY. The map assigning to each $p \in P(M \times_\rho T)$ the function $\chi \mapsto (p_\chi, a_\chi)$ cf. (3.1), (3.2) gives a bijection between $P(M \times_\rho T)$ and the set of all functions

$$\Sigma = X_m(T) \to (\mathrm{Der}_k M) \times (X_a(M)_1), \qquad \chi \mapsto (p_\chi, a_\chi)$$

satisfying the following conditions:

a) The map $\chi \mapsto a_\chi$ is a group homomorphism

b) $p_1 K \subset K$, $p_\chi \in L(M)_\chi$ for all $\chi \neq 1$ and for any $a \in \mathcal{O}(M)$ there are only finitely many χ's for which $p_\chi a \neq 0$.

c) The relations 2), 3), 4), 5) from (3.2) are satisfied (with $p^\circ = \Sigma p_\chi$).

Proof. Let's check first that if $p \in P(M \times_\rho T)$ then $p_\chi \in L(M)_\chi$ for $\chi \neq 1$. Indeed summing up the relations (3.2), 3) for all $\chi' \in \Sigma$ we get $\mu_M p_\chi = (p_\chi \otimes 1)\mu_m$ which means that $p_\chi \in L(M)$. Introducing the latter formula in (3.2), 3) again and applying $1 \otimes \varepsilon_M$ we get $f_{\chi'\chi^{-1}} p_\chi = p_\chi f_\chi$, which is what we wanted. To conclude it is sufficient to note that for any function $\chi \mapsto (p_\chi, a_\chi)$ satisfying a), b), c) above the formula

$$p(a \otimes \chi) = (\Sigma p_\chi a \otimes \chi'\chi) + a a_\chi \otimes \chi, \; a \in \mathcal{O}(M), \quad \chi \in \Sigma$$

defines an element $p \in P(M \times_\rho T)$.

(3.4) COROLLARY. Assume in notations above that $p \in P(M \times_\rho T)$. Then the following relations hold for p° and a_χ:

a) $\mu_M p^\circ = (p^\circ \otimes 1 + 1 \otimes p^\circ + \Sigma f_\chi \otimes a_\chi)\mu_M$

b) $p^\circ = S_M p^\circ S_M + \Sigma a_\chi f_\chi$

c) $p^\circ \varepsilon_M = \varepsilon_M p^\circ$

Proof. c) is (3.2), 4).

a) follows by summing up (3.2), 2) over all $\chi \in \Sigma$ and adding to the obtained relation the relations $\mu_M p_\chi = (p_\chi \otimes 1)\mu_M$ for $\chi \neq 1$ (cf. (3.3)).

b) follows by summing up (3.2), 5) over all $\chi \in \Sigma$.

(3.5) There are three remarkable maps from $P(M \times_\rho T)$ namely

$$P(M \times_\rho T) \to L(M), \quad p \mapsto p^+ := \sum_{\chi \neq 1} p_\chi$$

$$P(M \times_\rho T) \to W(M \times_\rho T), \quad p \to \ell\nabla_p, \quad \text{cf. (I. 1.4)}$$

$$P(M \times_\rho T) \to \mathrm{Der}_k M, \quad p \to p^\circ = p^+ + p_1, \quad \text{cf. (I. 1.4)}$$

Of course the second map assigns to any p the map $\chi \mapsto a_\chi$ cf. our previous notations in (3.3) and where we identify $W(M \times_\rho T)$ with $\mathrm{Hom}(\Sigma, X_a(M)_1)$. We have the following properties of these maps:

1) $p^o \in P(M) \Leftrightarrow p^o$ is S_M-invariant $\Leftrightarrow \ell\nabla_p$ vanishes on $\Phi(T,\mathcal{O}(M))$ (the latter simply means that p kills the weights, viewed as elements of $\mathcal{O}(M \times_\rho T)$; we shall repeatedly us this expression in what follows).

2) If $p^+ = 0$ then $p^o \in (\mathrm{Der}_k M)_1$

3) We have an exact sequence

$$0 \to P(M)_1 \to P(M \times_\rho T) \to L(M) \oplus W(M \times_\rho T)$$

4) We have

$$\{p \in P(M \times_\rho T); p^+ = 0\} \simeq \{(p^o, a) \in (\mathrm{Der}_k M)_1 \times \mathrm{Hom}(\Sigma, X_a(M)_1);$$
$$p^o, a \text{ satisfy a), b), c) in (3.4)}\}$$

5) The image of $\ell\nabla : P(M \times_\rho T) \to W(M \times_\rho T)$ contains the space of those maps $a : \Sigma \to X_a(M)_1$ which vanish on $\Phi(T,\mathcal{O}(M))$. So clearly equality holds provided $P(M \times_\rho T)$ kills the weights.

Proof. 1) follows directly from (3.4). To get 2) note that $p^o = p_1$ so introducing (3.4), a) in (3.2), 2) and appying $1 \otimes \varepsilon_M$ we get $f_\chi p^o = p^o f_\chi$ for all χwhich is what we want. To get 3) note that the inclusion $P(M)_1 \subset P(M \times_\rho T)$ is given by associating to any $p_1 \in P(M)_1$ the function $1 \to (p_1, 0)$, $\chi \to (0,0)$ for $\chi \neq 1$, cf. the identification in (3.3) hence clearly $P(M)_1$ is contained in the kernel of $P(M \times_\rho T) \to L(M) \oplus W(M \times_\rho T)$. The converse inclusion is also easily seen. Finally 4), 5) are easily checked through the identification (3.3): for 5) put $a_\chi = a(\chi)$ and $p_\chi = 0$ for all χ.

(3.6) REMARK. We will give at the end of (V.2) an example of a situation when $P(M \times_\rho T)$ contains a derivation not killing the weights!. This shows that $P(M \times_\rho T)$ can be rather complicated. In our example, T will be G_m (hence T will be algebraic) and M will be unipotent (but non-algebraic); we don't know whether it is possible to find such an example with both T and M algebraic. On the other hand we have the following:

(3.7) LEMMA. In notations above assume M is a vector group scheme (i.e. $\mathcal{O}(M)$ is the symmetric algebra over its subspace $X_a(M)$ of primitive elements). Then $P(M \times_\rho T)$ kills the weights of T in $\mathcal{O}(M)$. Equivalently, the image of the map $\ell\nabla : P(M \times_\rho T) \to W(M \times_\rho T) =$ $= \mathrm{Hom}(\Sigma, X_a(M)_1)$ is precisely the space of all maps in $\mathrm{Hom}(\Sigma, X_a(M)_1)$ vanishing on the set of weights $\Phi(T,\mathcal{O}(M))$.

Proof. Applying the twist map $\tau : \mathcal{O}(M) \otimes \mathcal{O}(M) \to \mathcal{O}(M) \otimes \mathcal{O}(M)$ to formula (3.4), a) and using the fact that $\tau\mu_M = \mu_M$ we get

$$\mu_M p^o = (1 \otimes p^o + p^o \otimes 1 + \Sigma a_\chi \otimes f_\chi)\mu_M$$

hence using (3.4), a) once again we get that

$$(\Sigma f_\chi \otimes a_\chi)\mu_M = (\Sigma a_\chi \otimes f_\chi)\mu_M$$

Let $x \in X_a(M)_\chi$, $x \neq 0$ be a primitive element of weight $\chi \neq 1$; applying the latter equality to x we get $x \otimes a_\chi = a_\chi \otimes \chi$. We claim that $a_\chi = 0$; for if it is not so the latter equality implies $x = \lambda a_\chi \in X_a(M)_1$ which is impossible since $\chi \neq 1$. So $a_\chi = 0$ for all weights of T in $X_a(M)$ (i.e. for all χ such that $X_a(M)_\chi \neq 0$). But the weights of T in $\mathcal{O}(M)$ are products of weights of T in $X_a(M)$ and hence $a_\chi = 0$ for all $\chi \in \Phi(T,\mathcal{O}(M))$.

(3.8) LEMMA. Assume L(M) contains a finite dimensional Lie subalgebra L with the following properties:

a) $L(M)_\chi \subset L$ for all $\chi \neq 1$ and $[P(M),L] \subset L$ in $\mathrm{Der}_k M$.

b) $\mathcal{O}(M)$ is locally nilpotent as an L-module.

c) $\mathcal{O}(M)$ is locally finite as a P(M)-module.

Put $P = \mathrm{Ker}(\ell\nabla : P(M \times_\rho T) \to W(M \times_\rho T))$. Then $\mathcal{O}(M \times_\rho T)$ is locally finite as a P-module; in particular $\mathrm{Ker}\,\ell\nabla \subset P(M \times_\rho T,\mathrm{fin})$.

Proof. First note that since $\dim_K L < \infty$ the set $\Phi^* = \{\chi \in \Sigma;\ L(M)_\chi \neq 0,\ \chi \neq 1\}$ of all non-trivial weights of T in L(M) is finite. Next we claim that for any $a \in \mathcal{O}(M)$, the K[P]-submodule of $\mathcal{O}(M) \otimes \mathcal{O}(T)$ generated by $a \otimes 1$ has finite dimension. Indeed, by (3.4) for $p \in P$ we have that $p^0 \in P(M)$ (because $a_\chi = 0$ for all χ) hence $p_1 = p^0 - p^+ \in P(M) \oplus L$ and $p_\chi \in L$ for $\chi \neq 1$. Since $\mathcal{O}(M)$ is a locally finite both as a P(M)-module and as an L-module and since $[P(M),L] \subset L$ it follows that $\mathcal{O}(M)$ is locally finite as a $P(M) \oplus L$-module. Choose a finite dimensional vector space V of $\mathcal{O}(M)$ containing $a \in \mathcal{O}(M)$ and stable under $P(M) \oplus L$. Hypothesis b) says that for V there exists an integer $N \geq 1$ such that $\theta_1\theta_2 \ldots \theta_N V = 0$ for all $\theta_1, \ldots, \theta_N \in L$. Then it is easy to check using $[P(M),L] \subset L$ once again that we have $\theta_1\theta_2 \ldots \theta_n V = 0$ for all $\theta_1, \ldots, \theta_n \in P(M) \oplus L$ provided card $\{i; \theta_i \in L\} \geq N$. Now let $\Phi^{(N)}$ be the K-span in $\mathcal{O}(T)$ of all products of the form $\chi_1\chi_2 \ldots \chi_N$ with $\chi_i \in \Phi^* \cup \{1\} = \Phi$. For $p^1, \ldots, p^n \in P$ we have

$$p^1p^2 \ldots p^n(a \otimes 1) = \Sigma p^1_{\chi_1} p^2_{\chi_2} \ldots p^n_{\chi_n} a \otimes \chi_n\chi_{n-1} \ldots \chi_1$$

the sum being taken for all n-uples $(\chi_1, \ldots, \chi_n)$ with $\chi_i \in \Phi$. As we have seen, $p^1_{\chi_1} p^2_{\chi_2} \ldots p^n_{\chi_n} a = 0$ whenever card $\{i; \chi_i \in \Phi^*\} \geq N$. Consequently $K[P](a \otimes 1) \subset V \otimes \Phi^{(N)}$ and our claim is proved. Now we claim (and this will close the proof) that for any $x \in \mathcal{O}(M) \otimes \mathcal{O}(T)$ we have $\dim_K K[P]x < \infty$. Indeed write $x = \Sigma a_i \otimes \chi_i$ and let V be a finite dimensional K[P]-submodule of $\mathcal{O}(M) \otimes \mathcal{O}(T)$ containing all a_i's. Then $\Sigma V(1 \otimes \chi_i)$ is a finite dimensional K[P]-module containing x and our lemma is proved.

(3.9) REMARK. If in (3.8) above we replace the condition c) by "c') $P(M) = P(M,\mathrm{fin})$ (in other words $\mathcal{O}(M)$ is locally finite as K[p]-module for any $p \in P(M)$)" then we still get the second part of the conclusion namely that "Ker $\ell \nabla \subset P(M \times_\rho T, fin)$".

This follows just by inspecting the proof of (3.8).

(3.10) LEMMA. Let G be an irreducible unipotent affine algebraic K-group. Then:

1) Assume char K > 0 and G is commutative. Then P(G) = P(G,fin).

2) Assume char K = 0. Then $\mathcal{O}(G)$ is locally finite as a D(G)-module.

Proof. 1) By [H] pp. 42 and 63-64, $\mathcal{O}(G)$ is locally nilpotent as an L(G)-module i.e. for all $y \in \mathcal{O}(G)$ there exists an integer N such that $\theta_1\theta_2 \dots \theta_N y = 0$ for all $\theta_1, \dots, \theta_N \in L(G)$. If $\lambda_x \in \mathrm{End}_K$ (G) denotes the multiplication in (G) by some element $x \in$ (G) then for any $\theta \in L(G)$ we have $[\theta, \lambda_x] = \lambda_{\theta x}$; so by (I. 1.3) if $x \in X_a(G)$ then $[\theta, \lambda_x]$ is the multiplication by some scalar in K. Now pick and element $p = \Sigma \lambda_{a_i} \theta_i = \Sigma a_i \otimes \theta_i \in X_a(G) \otimes L(G) = P(G/K) = P(G)$, cf. (I. 1.9) where $(\theta_i)_i$ is a K-basis of L(G) and $a_i \in X_a(G)$. Then it is easy to see using the above remarks that K[p]y is contained in the K-linear span in (G) of the set

$$\{\lambda_{a_{i_1} a_{i_2} \dots a_{i_n}} \theta_{j_1} \theta_{j_2} \dots \theta_{j_n} ; n \leq N\} \subset \mathcal{O}(G)$$

In particular $\dim_K K[p]y < \infty$ for all $y \in \mathcal{O}(G)$ and assertion 1) is proved.

2) By [H] p. 231 the image of the map (I. 1.13)

$$\mathcal{O}(G) \to \mathcal{O}_{G,e} \xrightarrow{j_G} U(L(G))^\circ$$

coincides with the algebra B(L(G)) of nilpotent representative functions on U(L(G)) (recall that by definition B(L(G)) consists of all functionals on U(L(G)) annihilating some power of the ideal J = L(G)U(L(G)). Now B(L(G)) is a locally finite D(G)-submodule of U(L(G)) since it is the union of the finite dimensional D(G)-submodules

$$B_n = \{f \in U(L(G))^\circ ;\ f \text{ vanishes on } J^n\}$$

Since j_G is a map of D(G)-modules, $\mathcal{O}(G)$ will be locally finite and we are done.

(3.11) REMARKS. 1) $\mathcal{O}(G)$ is not a locally finite D(G)-module even in the case $G = G_a$, char K = q > 0 (if $G_a = \mathrm{Spec}\, K[\xi]$, $p_i = \xi^{q^i} \frac{d}{d\xi} \in P(G)$ then $p_i \xi = \xi^{q^i}$ hence $\dim_K(P(G)\xi) = \infty$) so the conclusion in (3.10), 1) cannot be replaced by "$\mathcal{O}(G)$ is locally finite as a D(G)-module".

2) It would be interesting to know if in (3.10), 1) one can drop the commutativity assumption; one can formulate the conjecture that P(G) = P(G,fin) for any irreducible affine unipotent algebraic K-group G (char K arbitrary).

(3.12) Let G be an irreducible affine algebraic K-group, let T be a maximal torus of the radical of G and let $\Phi(T, \mathcal{O}(G)) \subset X_m(T)$ be the set of weights of the action of T on $\mathcal{O}(G)$ be inner automorphisms. Moreover let $\Phi(G)$ be the subset of $X_m(G)$ of all characters of G whose restriction to T belongs to $\Phi(T,\mathcal{O}(G))$; the elements of $\Phi(G)$ will be called the weights of G. Since the maximal tori of the radical are all conjugate, $\Phi(G)$ does not depend on the choice of T. We let $W_o(G)$ be the subspace of $W(G) = \mathrm{Hom}(X_m(G), X_a(G))$ consisting of all maps vanishing on $\Phi(G)$.

Here is our main result in arbitrary characteristic:

(3.13) COROLLARY. Let G be a solvable irreducible affine algebraic K-group.

1) Assume the unipotent radical of G is commutaive. Then the kernel of $\ell\nabla: P(G) \to W(G)$ is contained in P(G,fin).

2) Assume the unipotent radical of G is a vector group. Then P(G) kills the weights of G. So the image of $\ell\nabla: P(G) \to W(G)$ equals $W_o(G)$.

Proof. By standard structure theory [H], $G = M \rtimes T$ with M the unipotent radical of G and T a maximal torus. Applying Lemma (3.8) and Remark (3.9) for L = L(M) together with Lemma (3.10) we get assertion 1) of the Corollary. To check assertion 2) note that the weights of G as defined in (3.12) can be identified with the weights of T on $\mathcal{O}(M)$ as defined in (3.1) so we may conclude by (3.7).

(3.14) REMARK. In notations of (3.13) it is not true that P(G,fin) is contained in Ker $(\ell\nabla: P(G) \to W(G))$. Indeed take $G = G_a \times G_m$ as in (I. 3.9); then with notations from loc. cit. consider the derivation $p = (\xi - \xi^q)\frac{\partial}{\partial \xi} + \xi\chi\frac{\partial}{\partial \chi} \in P(G,fin)$ and note that $\ell\nabla_p \chi = \chi^{-1} p\chi = \xi \neq 0$. But this phenomenon does not occur in characteristic zero as shown by the following:

(3.15) LEMMA. Assume char K = 0 and G is an affine integral K-group scheme. Then

$$P(G,fin) = \mathrm{Ker}(\ell\nabla: P(G) \to W(G))$$

Proof. Let $p \in P(G,fin)$ and assume $\chi^{-1} p\chi = a \neq 0$ for some $\chi \in X_m(G)$. By (I. 1.3), $X_a(G)$ is stable under P(G). By [H] p. 88 the symmetric algebra $S(X_a(G))$ embeds into $\mathcal{O}(G)$ and then of course each homogeneous component $S^n(X_a(G))$ is a P(G)-submodule of $\mathcal{O}(G)$. By induction we get that

$$p^n\chi = (a^n + \xi_n)\chi \quad \text{for } n \geq 0$$

with $\xi_n \in \bigoplus_{i<n-1} S^i(X_a(G))$. By our assumption the K-span of the family $(p^n\chi)_n$ is finite dimensional; this implies that the same holds for the family $(a^n + \xi_n)_n$ which is impossible because $a^n \in S^n(X_a(G))$. The lemma is proved.

In what follows we devote ourselves to affine algebraic groups in characteristic zero.

(3.16) THEOREM. Let char K = 0 and let G be an irreducible affine algebraic K-group. Then:

1) $P(G,fin) = \mathrm{Ker}(\ell\nabla: P(G) \to W(G))$ and $\mathcal{O}(G)$ is locally finite as a P(G,fin)-module.

2) The image of $\ell\nabla : P(G) \to W(G)$ contains $W_o(G)$. More precisely there exists a K-linear map $E : W_o(G) \to P(G/K)$, $a \mapsto E_a$ such that $(\ell\nabla) \circ E$ is the natural inclusion $W_o(G) \subset W(G)$ and having the following properties :

a) Im E is an abelian ideal in $P(G/K,fin) \oplus \mathrm{Im}\, E (\subset P(G/K))$.

b) For any $\sigma \in \mathrm{Aut}\, G$ and $a \in W_o(G)$ we have $\sigma^{-1} E_a \sigma = E_{\sigma a}$ where we still denote by σ the induced automorphisms of $\mathcal{O}(G)$ and $W_o(G)$.

c) For any $p \in \mathrm{Der}(K/K_G)$ upon letting $G = G_o \otimes_{K_G} K$ (G_o a K_G-group) and letting p^* be the trivial liftings of p from K to $\mathcal{O}(G) = K \otimes_{K_G} \mathcal{O}(G_o)$ and to $W_o(G) = K \otimes_{K_G} W_o(G_o)$ we have $[p^*, E_a] = E_{p^* a}$ for all $a \in W_o(G)$.

REMARK. Assertion 1) can be easily deduced in fact from our previous result (I. 1.18), cf. (3.24) below; but our proof here for (3.16) will be purely algebraic.

We start with some preparations (cf. also [B_3]):

(3.17) LEMMA. Assume char K = 0 and g is a Lie D-algebra of finite dimension. Then the radical r of g is a Lie D-subalgebra.

Proof. By (I. 3.2) and (I. 3.4) g splits over some D-field extension K_1/K with $K_o := K_1^D = K^D$ so $g \otimes K_1 = g_o \otimes_{K_o} K_1$ for some Lie K_o-algebra g_o. Let r_o be the radical of g_o. Both $r_o \otimes_{K_o} K_1$ and $r \otimes_K K_1$ coincide with the radical of $g \otimes_K K_1$. But $r = (r \otimes_K K_1) \cap g = (r_o \otimes_{K_o} K_1) \cap g$ and the latter space is preserved by D.

(3.18) COROLLARY. Let char K = 0 and G be an irreducible affine algebraic K-group. Then the radical R of G is an algebraic D(G)-subgroup of G. In particular there is a natural Lie K/k-algebra restriction map $P(G) \to P(R)$.

Proof. Combine (I. 1.16) and (3.17) above.

(3.19) COROLLARY. Assume char K = 0 and M is an irreducible affine algebraic K-group whose radical is unipotent. Then $\mathcal{O}(M)$ is locally finite as a D(M)-module.

Proof. By (I. 3.2) and (I. 3.4) we may assume that $g = L(G)$ is a split Lie D-algebra so $g = g_o \otimes_{K_o} K$ ($K_o = K^D$, $g_o = g^D$, $D = D(G)$). Let $g_o = r_o + s_o$ be a decomposition of g_o with r_o its radical and s_o a complementary semisimple Lie algebra. Then by [H] p. 112, $s = s_o \otimes K$ is an algebraic Lie subalgebra of g, $s = L(S)$, $S \subset G$. By (I. 1.16) both S and the radical R of G are algebraic D-subgroups of G; therefore the multiplication map $R \times S \to G$ is a D-map hence $\mathcal{O}(G)$ identifies with a D-submodule of $\mathcal{O}(R) \otimes \mathcal{O}(S)$ so we are reduced to proving that both $\mathcal{O}(R)$ and $\mathcal{O}(S)$ are locally finite D-modules. The assertion about $\mathcal{O}(R)$ follows from (3.10) because R is unipotent. The assertion for $\mathcal{O}(S)$ follows for instance from (2.13) (it also follows from representability of <u>Aut</u> S [GD]; but the latter fact involves the whole structure theory of reductive group schemes over non-reduced basis so our argument (2.11) should be viewed as "the elementary" argument for the local finitness of $\mathcal{O}(S)$; for another "elementary" argument see $[B_3]$).

We shall repeatedly use the following remarks:

(3.20) Let $G = G_1 \rtimes G_2$ be a semidirect product of irreducible affine algebraic K-groups, char K = 0 and identify $\mathcal{O}(G)$ with $\mathcal{O}(G_1) \otimes \mathcal{O}(G_2)$ via the multiplication map $G_1 \times G_2 \to G$. If $X_m(G_1) = 1$ then $X_m(G)$ identifies with $X_m(G_2)$. If $X_a(G_2) = 0$ then $X_a(G) = X_a(G_1)^{G_2}$.

Now assume $G_2 \to G_1$ is an isogeny. Then the map $X_m(G_1) \to X_m(G_2)$ is injective with finite cokernel and the map $X_a(G_1) \to X_a(G_2)$ is an isomorphism. In particular there is an induced isomorphism $W(G_1) \simeq W(G_2)$.

(3.21) To prove Theorem (3.16) we fix some notations: let U be the unipotent radical of G, let H be a maximal reductive subgroup of G and T the radical of H. By a theorem of Mostow [H] p. 117, $G = U \rtimes H$; moreover T is of course a maximal torus of the radical R of G. Put $S = [H,H]$, $M = U \rtimes S$ and $\tilde{G} := M \rtimes T = U \rtimes (S \times T)$. The isogeny $S \times T \to H$ induces an isogeny $\tilde{G} \to G$. We write $\Sigma = X_m(T)$. Note that restriction map $X_m(G) \to \Sigma$ is injective and has finite cokernel; indeed this map is easily seen to identify with the map

$$X_m(G) \simeq X_m(H) \to X_m(S \times T) \simeq X_m(T)$$

cf. (3.20) above and we are done also by (3.20).

By (3.15) Ker $(\ell\nabla : P(G) \to W(G))$ contains P(G,fin); let's prove it actually equals P(G,fin). By (I. 1.21) and (3.20) we have a commutative square

$$\begin{array}{ccc} P(G) & \longrightarrow & W(G) \\ \downarrow & & \| \\ P(\tilde{G}) & \longrightarrow & W(\tilde{G}) \end{array}$$

so it is sufficient to prove that $\mathcal{O}(\tilde{G})$ is a locally finite P-module where $P = \mathrm{Ker}(\ell\nabla : P(\tilde{G}) \to W(\tilde{G}))$. Since T centralizes S, $L(M)_\chi = (L(U) \oplus L(S))_\chi = L(U)_\chi \subset L(U)$ for all $\chi \in \Sigma$, $\chi \neq 1$. Due to (3.19) we may apply Lemma (3.8) to our situation (with $L = L(U)$; note that $[P(M), L(U)] \subset L(U)$ by (3.18) and (I. 1.16)) so we get that $\mathcal{O}(\tilde{G})$ is locally finite as a P-module and assertion 1) in (3.16) is proved.

To prove assertion 2) let's define the map $E : W_0(G) \to P(G)$, $a \mapsto F_a$ as follows. The action of T on H by the left (or, which is the same by the right) translations gives a Σ-gradation

$$\mathcal{O}(H) = \bigoplus_\chi \mathcal{O}(H)^\chi$$

Let $f^\chi : \mathcal{O}(H) \to \mathcal{O}(H)^\chi \subset \mathcal{O}(H)$ be the corresponding projections. Restriction provides an identification $W(G) \simeq \mathrm{Hom}(\Sigma, X_a(U)^H)$. Then, any $a \in W_0(G) \subset \mathrm{Hom}(X_m(G), X_a(G))$ can be viewed as a homomorphism $\Sigma \to X_a(U)^H$ vanishing on the weights $\Phi(T,\mathcal{O}(G)) = \Phi(T,\mathcal{O}(M))$. Moreover, identify $\mathcal{O}(G)$ with $\mathcal{O}(U) \otimes \mathcal{O}(H)$ via the multiplication map $U \times H \to G$ and define the K-linear endomorphism E_a of $\mathcal{O}(G)$ by the formula:

$$E_a(x\otimes y) = \sum_\chi x\, a(\chi)\otimes f^\chi(y), \quad x\in\mathcal{O}(U),\ y\in\mathcal{O}(H).$$

Clearly E_a is an $\mathcal{O}(U)$-derivation. To check that $E_a \in P(G)$ it is sufficient to check that its lifting $\tilde{E}_a$ to $\mathcal{O}(\tilde{G}) = \mathcal{O}(U)\otimes\mathcal{O}(S\times T)$ belongs to $P(\tilde{G})$. But we have

$$\tilde{E}_a(x\otimes z) = \sum_\chi x a(\chi)\otimes\tilde{f}^\chi(y), \quad x\in\mathcal{O}(U), \quad z\in\mathcal{O}(S\times T)$$

where $\tilde{f}^\chi : \mathcal{O}(S\times T)\to\mathcal{O}(S\times T)^\chi\subset\mathcal{O}(S\times T)$ is the corresponding projection and $\mathcal{O}(S\times T)^\chi = \mathcal{O}(S)\otimes K\chi$. Now $\tilde{E}_a \in P(\tilde{G})$ by (3.3) (in notations of (3.3) put $a_\chi = a(\chi)$ and $p_\chi = 0$ for all χ).

So our map E is well-defined and clearly $(\ell\nabla)\circ E$ is the inclusion $W_o(G)\subset W(G)$ so $W_o(G)\subset \mathrm{Im}(\ell\nabla)$. It is also clear that $\mathrm{Im}(\ell\nabla)$ is an abelian subalgebra of $P(G/K)$. To prove the remaining assertions in claims a) and b) of (3.16) note that Aut G is generated by Int G and the group Aut(G,H) of all automorphisms of G preserving H. Consequently by (I. 3.12) $P(G/K,\mathrm{fin}) = L(\mathrm{Aut}\,G)$ is generated by L(Int G) and L(Aut(G,H)). So it is sufficient to check that the following hold for all $a\in W_o(G)$:

(3.21.1)	$\sigma^{-1}E_a\sigma = E_a$	for all $\sigma\in\mathrm{Im}(M\to\mathrm{Int}\,G)$
(3.21.2)	$[d,E_a] = 0$	for all $d\in\mathrm{Im}(L(M)\to L(\mathrm{Int}\,G))$
(3.21.3)	$\sigma^{-1}E_a\sigma = E_{\sigma a}$	for all $\sigma\in\mathrm{Aut}(G,H))$
(3.21.4)	$[d,E_a]\in\mathrm{Im}\,E$	for all $d\in L(\mathrm{Aut}(G,H))$

To prove the assertions above we may assume $G = \tilde{G}$. Start with (3.21.1) and (3.21.2). The action of M on G by inner automorphisms is given by the composition of maps

$$M\times M\times T \xrightarrow{a} M\times M\times T \xrightarrow{b} M\times M\times T \xrightarrow{c} M\times T$$

$$(m,x,t) \longmapsto (mx,m,t) \longmapsto (mx,tm^{-1}t^{-1},t) \mapsto (mxtm^{-1}t^{-1},t)$$

where a and c are induced by multiplications while b is induced by taking first the antipode on the middle factor and then applying the action $M\times T\to M$ of T on M by inner automorphisms. This immediately implies that for any $z\in\mathcal{O}(M)$ we have that the image of $z\otimes 1$ via the map

$$a^*b^*c^* : \mathcal{O}(M)\otimes\mathcal{O}(T)\to\mathcal{O}(M)\otimes\mathcal{O}(M)\otimes\mathcal{O}(T)$$

belongs to $\mathcal{O}(M)\otimes\mathcal{O}(M)\otimes\langle\Phi(G)\rangle$ where $\langle\Phi(G)\rangle\subset\mathcal{O}(T)$ is the K-span of the group generated by $\Phi(G) = \Phi(T,\mathcal{O}(G))$. This shows that if σ and d are as in (3.21.1) and (3.21.2) we have $\sigma(z\otimes 1)$, $d(z\otimes 1)\in\mathcal{O}(M)\otimes\langle\Phi(G)\rangle$. This immediately implies that the derivations $\sigma^{-1}E_a\sigma - E_a$ and $[d,E_a]$ vanish on $\mathcal{O}(M)\otimes 1$. On the other hand (since σ is the identity on $1\otimes\mathcal{O}(T)$ and on $X_a(G)$ and d is the zero map on these spaces), the above derivations also vanish on $1\otimes\mathcal{O}(T)$ hence they vanish on $\mathcal{O}(M)\otimes\mathcal{O}(T)$ and (3.21.1), (3.21.2) are proved. To prove (3.21.3) let σ still denote the restriction of σ to M and T. Then for $x \in \mathcal{O}(M)$, $\chi \in X_m(T) \subset \mathcal{O}(T)$ we have

$$\sigma^{-1}E_a\sigma(x \otimes \chi) = \sigma^{-1}E_a(\sigma x \otimes \sigma\chi) = \sigma^{-1}(a(\sigma\chi)\sigma x \otimes \sigma\chi) =$$
$$= \sigma^{-1}(a(\sigma\chi))x \otimes \chi = E_{\sigma a}(x \otimes \chi)$$

To prove (3.21.4) let d still denote the derivation induced on $\mathcal{O}(M)$; noting that d kills $X_m(T)$ we have for x and χ as above:

$$[d,E_a](x \otimes \chi) = xd(a(\chi)) \otimes \chi = E_{da}(x \otimes \chi)$$

So assertions a) and b) are proved. Assertion c) can be proved similarily.

(3.22) COROLLARY. Assume in (3.16) that either the unipotent radical of G is commutative or the radical of G is nilpotent. Then $\mathrm{Im}(\ell\nabla : P(G) \to W(G)) = W_o(G)$ and Im E is an abelian representative ideal in P(G). Moreover each element of Im E kills $X_a(G)$.

Proof. Assume the unipotent radical of G is commutative. By (3.18) we have a commutative diagram

$$\begin{array}{ccc} P(G) & \xrightarrow{\ \ell_G\nabla\ } & W(G) \\ \downarrow & & \downarrow \\ P(R) & \xrightarrow[\ \ell_R\nabla\ ]{} & W(R) \end{array}$$

Moreover the right vertical arrow is injective and viewing it as an inclusion we have $W(G) \cap W_o(R) = W_o(G)$. By (3.7) $\mathrm{Im}(\ell_R\nabla) \subset W_o(R)$ so we obtain that $\mathrm{Im}(\ell_G\nabla) \subset W_o(G)$. A similar argument shows that the latter holds if R is nilpotent. This and (3.16) show that $K^{P(G)} = K_G$ and $P(G) = P_1 \oplus P_2 \oplus P_3$ with $P_1 = \mathrm{Der}(K/K_G)$, $P_2 = P(G/K,\mathrm{fin})$, $P_3 = \mathrm{Im}\,E$ and

$$\begin{array}{lll} [P_1,P_1] \subset P_1, & [P_1,P_2] \subset P_2, & [P_1,P_3] \subset P_3 \\ & [P_2,P_2] \subset P_2, & [P_2,P_3] \subset P_3 \\ & & [P_3,P_3] = 0. \end{array}$$

This and the Aut G-invariance of Im E proved in (3.16) show that Im E is an abelian representative ideal. It clearly kills $X_a(G)$ so we are done.

EXAMPLE 1

Assume char K = 0, H is a reductive irreducible algebraic K-group with radical T, V is a K-linear space of finite dimension and let

$$\rho : H \to GL(V)$$

be a representation. Put $G = V x_\rho H$. Then (3.16) and (3.22) say that P(G) has an abelian representative ideal which naturally identifies with the space $W_0(G)$ of all homomorphisms $X_m(H) \to (V^o)^H$ which vanish on all those $\chi \in X_m(H)$ for which $\chi|_T$ belongs to the set $\Phi(T,V)$ of all weights of T in V .

So if Γ is the subgroup of $X_m(T)$ generated by $\Phi(T,V)$ then

$$\dim W_0(G) = (\operatorname{rank} H - \operatorname{rank} \Gamma)\dim((V^o)^H)$$

where recall that rank H = dim T, by definition. Consequently we get:

$$\dim L(\underline{\mathrm{Aut}}\, G) - \dim L(\mathrm{Aut}\, G) = (\operatorname{rank} H - \operatorname{rank} \Gamma)\dim((V^o)^H)$$

In particular $P(G) = P(G,\mathrm{fin})$ (equivalently $L(\mathrm{Aut}\, G) = L(\underline{\mathrm{Aut}}\, G)$) in each of the following cases

a) $\operatorname{rank} \Gamma = \operatorname{rank} H$

b) $(V^o)^H = 0$

EXAMPLE 2

We specialize the above example and then say more.

Start with a representation

$$\rho : GL_N \to GL(V)$$

and put $G = V \times_\rho GL_N$; then P(G) behaves quite differently according to whether ρ factors through PGL_N or not. Indeed:

2α) If ρ doesn't factor through PGL_N then $P(G) = P(G,\mathrm{fin})$ (indeed in notations of the previous example rank Γ = rank H = 1!).

2β) If ρ factors through PGL_N then P(G) has an abelian representative ideal of dimension equal to the dimension of the space $(V^o)^{GL_N}$ of fixed elements of ρ in V^o. We may describe this ideal explicitely as follows. Let $G = \operatorname{Spec} K[\xi, y, \frac{1}{\det y}]$, $\xi = (\xi_1, \dots, \xi_n)$, $y = (y_{ij})_{1 \le i,j \le N}$ where ξ is a basis of V^o. Then $\mathfrak{J}$ consists of all derivations of the form:

$$\rho_f = f(\xi)\Big(\sum_{i,j} y_{ij}\frac{\partial}{\partial y_{ij}}\Big), \quad f = f(\xi) \in (V^o)^{GL_N} \subset \Sigma K\xi_i$$

By (3.16) we have seen that $W_0(G) \subset \mathrm{Im}(\ell\nabla)$. In what follows we use the (analitically proved) result (1.1) to "aproximate $\mathrm{Im}(\ell\nabla)$ from above" more precisely to show that $\mathrm{Im}(\ell\nabla)$ is contained in the kernel of a natural (Lie algebra theoretically defined) map $W(G) \to H^2(L(U),L(U))$.

So fix an irreducible affine algebraic K-group G (char K = 0) with radical R and unipotent radical U, pick a maximal torus T of R and put r = L(R), u = L(U), t = L(T).

(3.23) We start with the remark that W(G) naturally identifies with $\mathrm{Hom}(u/[g,g] \cap u, r/u)$. Indeed, identify both $L(G_m)$ and $L(G_a)$ with the field K via the identifications $G_a = \operatorname{Spec} K[t]$, $G_m = \operatorname{Spec} K[t,t^{-1}]$. Moreover write $G/[G,G] = G^{(a)} \times G^{(m)}$ with $G^{(a)}$ a vector group and $G^{(m)}$ a torus; clearly the natural map $U/U \cap [G,G] \to G^{(a)}$ is an isomorphism while the map $T \to G^{(m)}$ is an isogeny. We have identifications $L(G^{(a)})^o \simeq Hom(L(G^{(a)}), L(G_a)) \simeq Hom(G^{(a)}, G_a) = X_a(G)$ and $L(G^{(m)})^o = Hom(L(G^{(m)}), L(G_m)) = X_m(G) \otimes K$. Hence we have an identification:

$$W(G) = Hom(X_m(G) \otimes K, X_a(G)) = Hom(L(G^{(m)})^o, L(G^{(a)})^o) =$$

$$= Hom(L(G^{(a)}), L(G^{(m)})) = Hom(u/u \cap [g,g], r/u)$$

(3.24) Next let's note that for $p \in P(G)$ the image of $\ell\nabla_p$ in $Hom(u/u \cap [g,g], r/u)$ (still denoted so) has the following particularly simple description: upon letting $\Pi : r \to r/u$ be the natural projection we claim that

$$\ell\nabla_p \hat{x} = \Pi(px) \quad \text{for all } x \in u$$

where $\hat{x}$ denotes the image of x in $u/u \cap [g,g]$. In particular $\ell\nabla_p = 0$ if and only if $pu \subset u$ (this together with (3.16) and (I. 1.16) provide a purely algebraic proof of (1.18)). To check our claim note that under the identification in (3.23) we have the following formula

$$(d\chi) \circ \ell\nabla_p = d(\chi^{-1}p\chi)$$

for all character $\chi : R/U \to G_m$ (where $d\chi : L(R/U) \to L(G_m) = K$ is the tangent map of χ, $\chi^{-1}p\chi \in X_a(G)$ is viewed as an additive character $U/U \cap [G,G] \to G_a$ and similarily $d(\chi^{-1}p\chi)$ is its tangent map). Identifying now $L(U/U \cap [G,G])$ (respectively $L(R/U)$) with a subspace of $\mathcal{O}(U/U \cap [G,G])^o$ (respectively $\mathcal{O}(R/U)^o$) the above formula reads

$$(\ell\nabla_p \hat{x})(\chi) = \hat{x}(\chi^{-1}p\chi)$$

for all $x \in L(U)$. But $\hat{x}(\chi^{-1}p\chi)$ coincides with the image of χ under the map

$$\mathcal{O}(R/U) \to \mathcal{O}(R) = \mathcal{O}(U) \otimes \mathcal{O}(T) \xrightarrow{p} \mathcal{O}(U) \otimes \mathcal{O}(T) \xrightarrow{1 \otimes \varepsilon_T} \mathcal{O}(U) \xrightarrow{x} K$$

and our claim is proved.

(3.25) Now we have the following Lie algebra theoretic construction. Let r be any Lie K-algebra, u an ideal in r containing $[r,r]$ and $s : r/u \to r$ a Lie algebra section of the projection $r \to r/u$. For each $f \in Hom(u/[r,r], r/u)$, the bilinear alternating form $b(f) : u \times u \to u$ defined by

$$b(f)(x,y) = [sf\hat{x}, y] - [sf\hat{y}, x] \quad \text{for } x,y \in u$$

(where $\hat{x}, \hat{y}$ are the images of x,y in $u/[r,r]$) is easily seen to be a 2-cocycle of u in u so we can consider the linear map induced by b:

$$\beta : Hom(u/[r,r], r/u) \to H^2(u,u)$$

(3.26) Coming back to our specific situation where $u = L(U)$, $r = L(R)$ and noting that

$Hom(u/u \cap [g,g], r/u)$ is a subspace of $Hom(u/[r,r], r/u)$ we get a map (which we still call β) $\beta : W(G) \to H^2(u,u)$ by composing the map β from (3.25) with the identification isomorphism $W(G) \simeq Hom(u/u \cap [g,g], r/u)$ from (3.23). Then we have:

(3.27) PROPOSITION. Let char K = 0 and G be an irreducible affine algebraic K-group. Then there is a complex exact in the first two terms:

$$0 \to P(G,fin) \to P(G) \xrightarrow{\ell\nabla} W(G) \xrightarrow{\beta} H^2(L(U),L(U))$$

where U is the unipotent radical of G.

Proof. By (3.16) the only thing to prove is that $Im(\ell\nabla) \subset Ker\,\beta$. So assume $a = \ell\nabla_p \in W(G)$ for some $p \in P(G)$. By (1.1) $K_0 = K^{K[p]}$ is a field of definition for G. Writing $G = G_0 \otimes_{K_0} K$ (G_0 a K_0-group) let's denote by p^* the trivial lifting of p from K to G. Moreover let $e_1 : r \to r$, $e_2 : r \to r$ be the projections onto u and t respectively. Then by (3.24), (3.25) we have

(*) $\qquad b(\ell\nabla_p)(x,y) = [e_2px,y] - [e_2py,x] \quad$ for all $x,y \in u$.

Projecting the equality $p[x,y] = [px,y] + [x,py]$ on u and using the relation (*) we get

(**) $\qquad e_1p[x,y] = [e_1px,y] + [x,e_1py] + b(\ell\nabla_p)(x,y)$

Now clearly p^* maps u into u and we have

(***) $\qquad p^*[x,y] = [p^*x,y] + [x,p^*y]$

Substracting (***) from (**) and putting $v = e_1p - p^* \in End_K u$ we get

$$b(\ell\nabla_p)(x,y) = v[x,y] - [vx,y] - [x,vy]$$

which shows that $b(\ell\nabla_p)$ is a coboundary and we are done.

(3.28) REMARK. Note that in the abstract frame (3.25) one can easily check that for any $f \in Hom(u/[r,r], r/u)$ the form b(f) defines in fact a (new) Lie algebra multiplication on u! Coming back to our specific situation (3.26) we see that for any $p \in P(G)$ we get a "new Lie algebra multiplication" $b(\ell\nabla_p) : u \times u \to u$ on u. It would be interesting to understand what information about p carries this "new Lie algebra multiplication".

We end this section by recording some more consequences of our results (for a generalisation to the non-linear case, see (IV. I.2)):

(3.29) COROLLARY. Let char K = 0 and G be an irreducible affine algebraic K-group with radical R. Then:

1) The natural restriction map $P(G)/P(G,fin) \to P(R)/P(R,fin)$ is injective.

2) If $i : G \to G'$ is an isogeny and $i^* : P(G') \to P(G)$ is the natural lifting map, then

$i^*(P(G',fin)) = P(G,fin) \cap i^*(P(G'))$.

3) If the center of G is finite then $P(G) = P(G,fin)$.

4) If $X_m(G) = 1$ or $X_a(G) = 0$ then $P(G) = P(G,fin)$.

Proof. 1) follows from (3.16) and the fact that the map $W(G) \to W(R)$ is injective, cf. (3.21). 4) follows from (3.16). 2) follows from (3.16) and the fact that the map $W(G') \to W(G)$ is an isomorphism, cf. (3.20).

To check 3) start with a preparation. Assume V is an N-dimensional D-module. Then the coordinate algebra $\mathcal{O}(GL(V))$ of $GL(V)$ has a natural structure of Hopf D-algebra defined by identifying $\mathcal{O}(GL(V))$ with $S(gl(V)^0)[d^{-1}]$ where S = "symmetric algebra" and $d \in S^N(gl(V)^0)$ is the "determinant". We claim that $\mathcal{O}(GL(V))$ is locally finite; indeed $S(gl(V)^0)$ clearly is so and we are done by noting that d is killed by P (to check this replace K by some D-field extension of it such that V splits so V will have a K-basis contained in V^D. Associated to this basis there is a P-constant basis X_{ij} of $gl(V)$; now d is a polynomial in the X_{ij}'s with integer coefficients so is killed by P!).

Coming back to our group G, let $Ad : G \to GL(g)$, $g = L(G)$ be its adjoint representation. Using the description of Ad in [H] p. 51 one checks that $Ad^* : \mathcal{O}(GL(g)) \to \mathcal{O}(G)$ is a D(G)-algebra map. Consequently if Z is the center of G, $\mathcal{O}(G/Z)$ is a locally finite D(G)-module (being identified with $\mathcal{O}(GL(g))/\mathrm{Ker}\, Ad^*$). By assertion 2), $\mathcal{O}(G)$ must be locally finite as a K[p]-module for all $p \in P(G)$ and we are done.

EXAMPLE

The example 2α) in (3.22) can be arranged to provide an example of affine algebraic K-group G for which:

i) $P(G) = P(G,fin)$.

ii) G has a positive dimensional center.

iii) $X_m(G) \neq 1$.

iv) $X_a(G) \neq 0$.

Indeed it is sufficient to choose $\rho = \rho_1 \oplus \rho_2$ where

$$\rho_1 : GL_N \to GL(V_1) \text{ is trivial, } \dim V_1 > 0$$

$$\rho_2 : GL_N \to GL(V_2) \text{ does not factor through } PGL_N .$$

We can rephrase part of our results in terms of algebraic D-groups:

(3.30) COROLLARY. Let G be an irreducible affine algebraic D-group (char K arbitrary, D = K[P]). Then

1) Assume dim P = 1, G solvable and the unipotent radical of G is commutative. If the

group like elements of $\mathcal{O}(G)$ are P-constants then G is locally finite.

2) Assume char K = 0. Then the radical R of G is an algebraic D-subgroup of G. Moreover G is locally finite if and only if R is so, if and only if all group like elements of $\mathcal{O}(G)$ are P-constants.

3) Assume char K = 0 and $i : G \to G'$ is an isogeny of algebraic D-groups. Then G is locally finite if and only if G' is so.

4) Assume char K = 0 and the center of G is finite. Then G is locally finite.

5) Assume char K = 0 and either $X_m(G) = 1$ or $X_a(G) = 0$. Then G is locally finite.

CHAPTER 3. COMMUTATIVE ALGEBRAIC D-GROUPS

In the first section of this chapter we introduce the "logarithmic Gauss-Manin connection" and the "total Gauss-Manin connection" associated to it. In the second section we prove a "duality theorem" saying that the "total Gauss-Manin connection" on the "total de Rham space" $H^1_{DR}(A)_t$ of an abelian variety A is isomorphic as a D-module with the continuous dual of the inverse limit of the Lie algebras of the "relative projective hulls" of A, viewed with its "adjoint" D-module structure. This implies a precise description of P(C) for any irreducible commutative algebraic K-group C; note that $K^{P(C)}$ need not be a field of definition for C! We also get from the above duality the following regularity theorem: let C be an irreducible algebraic D-group with $\mathcal{O}(C) = K$ ($D = K[P]$, $P = \mathrm{Der}_k K$, tr. deg. $K/k < \infty$); then L(C) is a regular D-module in Deligne's sense. This together with a "descent criterion" will be proved in section 3.

Everywhere in this chapter K is algebraically closed of characteristic zero and if not otherwise specified P is any Lie K/k-algebra and D = K[P].

1. Logarithmic Gauss-Manin connection

Let V be a smooth projective K-variety. Recall from [Ka], [K0] that there is a natural integrable connection

$$\nabla : \mathrm{Der}_k K \to \mathrm{Hom}_k(H^1_{DR}(V), H^1_{DR}(V)), \quad p \mapsto \nabla_p$$

called the "Gauss-Manin" connection; here $H^1_{DR}(V)$ is the first de Rham cohomology space of V (see (1.1) for a quick review of this). Recall from [MM] that there is a "multiplicative analogue" of $H^1_{DR}(V)_a := H^1_{DR}(V)$ which is an abelian group called here $H^1_{DR}(V)_m$, see (1.2); we shall define a K-linear map

$$\ell\nabla : \mathrm{Der}_k K \to \mathrm{Hom}_{gr}(H^1_{DR}(V)_m, H^1_{DR}(V)_a)$$

We shall call the pair $(\nabla, \ell\nabla)$ the logarithmic Gauss-Manin $\mathrm{Der}_k K$-connection on $(H^1_{DR}(V)_m, H^1_{DR}(V)_a)$; $H^1_{DR}(V)_a$ will be viewed as an abelian Lie K-algebra on which $H^1_{DR}(V)_m$ acts trivially.

(1.1) For convenience of the reader and also for computational purposes we recall briefly the construction of ∇ cf. [K0]. Let $\mathcal{U}=(U_i)_i$ be an affine covering of V, let $C_a^{\bullet\bullet}=(C_a^{ts})$,

$$C_a^{ts}=C_a^{ts}(\mathcal{U})=\bigoplus_{i_0<i_1\dots<i_s} H^0(U_{i_0\dots i_s},\Omega^t_{V/K})$$

be the standard "Čech-de Rham" double complex of V and let $C_a^{\bullet}=C_a^{\bullet}(\mathcal{U})$ be the simple complex associated to $C_a^{\bullet\bullet}$. Recall that the differential $C_a^0\to C_a^1$ is given by the formula

$$(f_i)\to(df_i,f_j-f_i),\quad f_i\in H^0(U_i,\mathcal{O}_V).$$

while the differential $C_a^1\to C_a^2$ is given by the formula

$$(\omega_i,x_{ij})\to(d\omega_i,\omega_j-\omega_i-dx_{ij},x_{jk}-x_{ik}+x_{ij})$$

where $\omega_i\in H^0(U_i,\Omega^1_{V/K})$, $x_{ij}\in H^0(U_i\cap U_j,\mathcal{O}_V)$.
By definition, the 1-st de Rham cohomology space $H^1_{DR}(V)$ (called here also $H^1_{DR}(V)_a$) is the 1-st hypercohomology K-linear space of the de Rham complex (called $DR(V)_a$):

$$0\to\mathcal{O}_V\xrightarrow{d}\Omega^1_{V/K}\xrightarrow{d}\Omega^2_{V/K}\xrightarrow{d}\dots$$

Recall that $H^1_{DR}(V)_a=H^1(C_a^{\bullet})$. Recall also from [Del$_1$] that the spectral sequence $E_r^{ts}(a)$ associated to $C_a^{\bullet\bullet}$ degenerates in E_1. In particular we have an exact sequence:

(1.1.1) $$0\to H^o(\Omega^1_{V/K})\to H^1_{DR}(V)_a\xrightarrow{\Pi_a}H^1(\mathcal{O}_V)\to 0$$

The connection ∇ is defined as follows. Let $\theta_i\in \mathrm{Der}_k(U_i)$ be any lifting of $p\in\mathrm{Der}_k K$. Then ∇_p takes a class in $H^1_{DR}(V)_a$ represented by $(\omega_i,x_{ij})\in C_a^1$ into the class represented by

$$(\mathrm{Lie}_{\theta_i}\omega_i,(\theta_j-\theta_i)\wedge\omega_j+\theta_i x_{ij})\in C_a^1$$

Here $\mathrm{Lie}_{\theta_i}:H^0(U_i,\Omega^1_{V/K})\to H^0(U_i,\Omega^1_{V/K})$ is the "Lie derivative" in the direction θ_i (recall that $H^0(U_i,\Omega^1_{V/K})=J/J^2$ where $J=\mathrm{Ker}(\mathcal{O}(U_i)\otimes\mathcal{O}(U_i)\to\mathcal{O}(U_i))$ so $H^0(U_i,\Omega^1_{V/K})$ has a natural structure of $K[\theta_i]$-module; then Lie_{θ_i} is by definition the multiplication with θ_i in this module structure); moreover $\wedge$ is the contraction between K-derivations and 1-forms with values regular functions.

(1.2) Consider now the complex $DR(V)_m$ of abelian sheaves [MM]:

$$1\to\mathcal{O}_V^*\xrightarrow{\mathrm{dlog}}\Omega^1_{V/K}\xrightarrow{d}\Omega^2_{V/K}\xrightarrow{d}\dots$$

and let $H^1_{DR}(V)_m$ denote the 1-st hypercohomology group of this complex cf. for instance [MM] p. 31. So if $C_m^{\bullet\bullet}$ is the direct limit (over all coverings $\mathcal{U}$) of the double complexes $C_m^{\bullet\bullet}(\mathcal{U})$ associated to $DR(V)_m$ and $\mathcal{U}$ and if $C_m^{\bullet}$ is the simple complex associated to the double complex

$C^{\bullet\bullet}_m$ then we have

$$H^1_{DR}(V)_m = H^1(C^\bullet_m)$$

Now let $E_r^{ts}(m)$ be the spectral sequence associated to $C^{\bullet\bullet}_m$; it does not degenerate in E_1 like in the "additive case" but one can easily check that

$$E_1^{01}(m) = \mathrm{Pic}(V) \quad \text{and}$$

$$E_1^{ts}(m) = H^s(\Omega^t_{V/K}) \quad \text{for } t \geq 1,\ s \geq 0$$

Consequently

$$E_2^{01}(m) = \mathrm{Pic}^\tau(V) \quad \text{(see [G])}$$

and by degeneration in E_1 of $E_r^{ts}(a)$,

$$E_2^{20}(m) = H^0(\Omega^2_{V/K})$$

It is known (essentially by [MM]) that

$$d_2^{01}(m) : E_2^{01}(m) \to E_2^{20}(m)$$

is the zero map. For convenience we check this; it is sufficient to check that $d_2^{01}(m)$ is a morphism of algebraic groups (because $\mathrm{Pic}^\tau(V)$ is an extension of a finite group by an abelian variety while $H^0(\Omega^2_{V/K})$ is an algebraic vector group so no algebraic nonzero homomorphism can exist between them). To check algebraicity of $d_2^{01}(m)$ consider for any noetherian reduced K-algebra R the complex

$$1 \to \mathcal{O}^*_{V\otimes R} \xrightarrow{\text{dlog}} \Omega^1_{V\otimes R/R} \xrightarrow{d} \Omega^2_{V\otimes R/R} \xrightarrow{d} \cdots$$

Let $C^{ts}(m)_R$ be the associated double complex and $E_r^{ts}(m)_R$ the corresponding spectral sequence. Exactly as above we have:

$$E_1^{01}(m)_R = \mathrm{Pic}(V\otimes R)$$

$$E_1^{ts}(m)_R = H^s(\Omega^t_{V\otimes R/R}) = H^s(\Omega^t_{V/K})(R)$$

(where we identified $H^s(\Omega^t_{V/K})$ with its associated algebraic vector group $\mathrm{Spec}(S(H^s(\Omega^t_{V/K})))$. Clearly

$$d_1^{01}(m)_R : \mathrm{Pic}(V\otimes R) \to H^1(\Omega^1_{V\otimes R/R})$$

factors through $\mathrm{Pic}(V\otimes R)/\mathrm{Pic}\,R = \mathrm{Pic}_{V/K}(R) = \{$R-points of $\mathrm{Pic}_{V/K}\}$. Moreover, since R is reduced the kernel of $\mathrm{Pic}_{V/K}(R) \to H^1(\Omega^1_{V\otimes R/R})$ equals $\mathrm{Pic}^\tau_{V/K}(R)$. Analogously we claim that

$$d_2^{01}(m)_R : E_2^{01}(m)_R \to E_2^{20}(m)_R = H^o(\Omega^2_{V\otimes R/R})$$

factors through $E_2^{01}(m)_R/\mathrm{Pic}\,R = \mathrm{Pic}^\tau_{V/K}(R)$. Indeed $E_2^{01}(m)_R$ is a quotient of

$$\{a \in C^{10}(m)_R + C^{01}(m)_R;\ da \in C^{20}(m)_R\}$$

and $d_2^{01}(m)_R$ is defined by taking the class of a into the class of da. Now if $a = (\omega_i, y_{ij})$ then $da = (d\omega_i,\ \omega_j - \omega_i - y_{ij}^{-1}dy_{ij},\ y_{jk}y_{ik}^{-1}y_{ij})$ so (y_{ij}) is a cocycle and $\omega_j - \omega_i = y_{ij}^{-1}dy_{ij}$. So if a comes from Pic R we have (after suitably refining the covering) that $y_{ij} = u_i u_j^{-1} v_{ij}$ for suitable u_i's with v_{ij} belonging to some ring of quotients of R; we get

$$\omega_j - \omega_i = u_i^{-1}du_i - u_j^{-1}du_j$$

(because $dv_{ij} = 0$) hence

$$d\omega_i = d(\omega_i + u_i^{-1}du_i)$$

But $\omega_i + u_i^{-1}du_i$ patch together to give a global form so the class of da vanishes and our claim is proved.

We get group homomorphisms

$$\bar{d}_2^{01}(m)_R : \mathrm{Pic}^\tau_{V/K}(R) \to (H^o(\Omega^2_{V/K}))(R)$$

behaving functorially in R for all R reduced, hence we get a morphism of algebraic K-groups $\mathrm{Pic}^\tau_{V/K} \to H^o(\Omega^2_{V/K})$ (because both these groups are reduced). So the claim that $d_2^{01}(m) = 0$ follows. In particular we get an exact sequence

$$(1.2.1) \qquad 0 \to H^o(\Omega^1_{V/K}) \to H^1_{DR}(V)_m \xrightarrow{\Pi_m} \mathrm{Pic}^\tau(V) \to 0$$

Now we define $\ell\nabla_p \in \mathrm{Hom}_{gr}(H^1_{DR}(V)_m,\ H^1_{DR}(V)_a)$ for $p \in \mathrm{Der}_k K$; let $\mathcal{U} = (U_i)$ be an affine covering of V, let $\theta_i \in \mathrm{Der}_k \mathcal{O}(U_i)$ be liftings of p, let $C^\bullet_m(\mathcal{U})$ be the simple complex associated to the double complex $C^{\bullet\bullet}_m(\mathcal{U})$ and define a group homomorphism $C^1_m(\mathcal{U}) \to C^2_m(\mathcal{U})$ by the formula

$$(\omega_i, y_{ij}) \to (\mathrm{Lie}_{\theta_i}\omega_i, (\theta_j - \theta_i) \,\overline{\wedge}\, \omega_j + y_{ij}^{-1}\theta_i y_{ij})$$

$\omega_i \in H^0(U_i, \Omega^1_{V/K})$, $y_{ij} \in \mathcal{O}^*(U_i \cap U_j)$.

One checks by hand computation that this morphism is well-defined, passes to cohomology and agrees with "refining " so we get a group homomorphism

$$\ell\nabla_p : H^1_{DR}(V)_m \to H^1_{DR}(V)_a$$

which depends K-linearly on $p \in P$.

The pair $(\nabla, \ell\nabla)$ defines a logarithmic $\mathrm{Der}_k K$-connection on $(H^1_{DR}(V)_m, H^1_{DR}(V)_a)$ called in what follows the logarithmic Gauss-Manin connection. We will prove later that it is integrable for V an abelian variety (i.e. that (0.16.2) holds).

(1.3) REMARK. Recall that if $K = \mathbf{C}$ then $H^1_{DR}(V) \simeq H^1(V^{an}, \mathbf{C})$ and $H^1_{DR}(V)_m \simeq H^1(V^{an}, \mathbf{C}^*)$. Moreover these two groups can be interpreted in terms of differentials of the second respectively of the third kind on V, cf. for instance: N. Katz, Invent. Math. 18, 1-2 (1972) p. 108. We won't need this interpretation in what follows.

(1.4) REMARK. From the very definitions we have (in notations of (1.1) and (1.2)) that for any $p \in \mathrm{Der}_k K$ the following diagram is commutative

$$\begin{array}{ccc} H^0(\Omega^1_{V/K}) & \hookrightarrow & H^1_{DR}(V)_m \\ \cap\downarrow & & \downarrow \ell\nabla_p \\ H^1_{DR}(V)_a & \xrightarrow[\nabla_p]{} & H^1_{DR}(V)_a \end{array}$$

(1.5) REMARK. It follows from [K0], [Ka] (and indeed from the very definitions) that for any $p \in P$ the K-linear map

$$H^0(\Omega^1_{V/K}) \to H^1_{DR}(V)_a \xrightarrow{\nabla_p} H^1_{DR}(V)_a \xrightarrow{\Pi_a} H^1(\mathcal{O}_A)$$

coincides with the cup product with the Kodaira-Spencer class $\rho(p) \in H^1(T_A)$ (where $\rho : \mathrm{Der}_k K \to H^1(T_A)$ denotes the Kodaira-Spencer map).

(1.6) In proving integrability of the logarithmic Gauss-Manin connection (and also in formulating our "duality theorem" in the next section) we are led to make a certain construction of linear algebra whose abstract version we now describe.

Let's consider the following data: an abelian group H_m, a K-linear space H_a and a K-linear space H_o equipped with a group homomorphism $j_m : H_o \to H_m$ and K-linear map $j_a : H_o \to H_a$ (both j_m, j_a injective). By a total space for these data we will mean a K-linear space H_t equipped with a group homomorphism $i_m : H_m \to H_t$ and with a K-linear map $i_a : H_a \to H_t$ such that the following diagram is commutative

$$\begin{array}{ccc} H_o & \xrightarrow{j_m} & H_m \\ j_a\downarrow & & \downarrow i_m \\ H_a & \xrightarrow[i_a]{} & H_t \end{array}$$

and such that for any other K-linear space H', any group homomorphism $i'_m : H_m \to H'$ and any K-linear map $i'_a : H_a \to H'$ with $i'_m \circ j_m = i'_a \circ j_a$ there is a unique K-linear map $f : H_t \to H'$ such that $f \circ i_m = i'_m$ and $f \circ i_a = i'_a$. Of course (H_t, i_m, i_a) is then unique up to a canonical isomorphism; it is a sort of "fibred sum" of H_m and H_a over H_o.

A total space always exists; it is constructed as follows. First put

$$H_{m,K} = (H_m \otimes_{\mathbf{Z}} K)/\mathrm{Ker}(H_o \otimes_{\mathbf{Z}} K \to H_o \otimes_K K = H_o)$$

and note that we have an exact sequence of K-linear spaces:

$$0 \to H_o \to H_{m,K} \to (H_m/H_o) \otimes K \to 0$$

Then define H_t as the cokernel of the map δ below

$$0 \to H_o \xrightarrow{\delta} H_{m,K} \oplus H_a \to H_t \to 0$$

where $\delta(x) = (x,-x)$ (we view j_m, j_a as inclusions!).

Note that i_a is always injective and we have

$$\dim_K H_t = \dim_K H_a + \mathrm{rank}(H_m/H_o)$$

(1.7) Now assume in situation above that we dispose of a logarithmic P-connection $(\nabla, \ell\nabla)$ on (H_m, H_a), where H_a is viewed as an abelian Lie algebra on which H_m acts trivially, such that for any $p \in P$ we have a commutative diagram

$$\begin{array}{ccc} H_o & \xrightarrow{j_m} & H_m \\ {\scriptstyle j_a}\big\downarrow & & \big\downarrow{\scriptstyle \ell\nabla_p} \\ H_a & \xrightarrow[\nabla_p]{} & H_a \end{array}$$

Then we can define in a canonical way a P-connection on the K-linear space $(H_m \otimes K) \oplus H_a$ by the formula

$$\nabla_p(y \otimes \lambda, x) = (y \otimes \partial(p)(\lambda), (\ell\nabla_p y) \otimes \lambda + \nabla_p x), \quad p \in P,\ y \in H_m,\ x \in H_a,\ \lambda \in K\ .$$

This P-connection induces a P-connection $\nabla : P \to \mathrm{Hom}_k(H_t, H_t)$ which will be called the total connection associated to our logarithmic connection (relative to H_o !). The basic (trivial) fact about this construction is the following:

(1.8) LEMMA. In notations above, the logarithmic P-connection $(\nabla, \ell\nabla)$ on (H_m, H_a) is integrable if and only if the associated total connection ∇ on H_t is integrable.

(1.9) Coming back to our logarithmic Gauss-Manin connection (with $H_m = H^1_{DR}(V)_m$, $H_a = H^1_{DR}(V)_a$, $H_o = H^0(\Omega^1_{V/K})$, see Remark (1.4)) we may form the total de Rham space $H_t = H^1_{DR}(V)_t$ viewed with the total Gauss-Manin connection on it cf. (1.7). This object will play a key role in the next section.

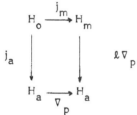

2. Duality theorem

(2.1) Let C be an irreducible commutative algebraic K-group, let B its maximum linear connected subgroup (which will be called the linear part of C) and consider $A = C/B$ (which will be called the abelian part of C). By [Se] there are naturally associated maps

$$m : X_m(B) \to \mathrm{Pic}^0(A)$$
$$a : X_a(B) \to H^1(\mathcal{O}_A)$$

with m a group homomorphism and a a K-linear map.
(Recall that they are defined as follows: one takes a covering $(U_i)_i$ of A and sections $s_i : U_i \to C$ of the projection $C \to A$; then for $\chi \in X_m(B)$ and $\xi \in X_a(B)$ we let $m(\chi) \in \mathrm{Pic}^0(A)$ be represented by the cocycle $\chi \circ (s_j - s_i)$ and $a(\xi) \in H^1(\mathcal{O}_A)$ be represented by the cocycle $\xi \circ (s_j - s_i)$. Recall also that the maps (m,a) uniquely determine the extension class of C as an extension of A by B. Finally recall that given an abelian variety A, an irreducible commutative algebraic group B and maps (m,a) as above there exists an extension C of A by B whose "associated" maps are (m,a). Consider $S(C)_m = \mathrm{Im}(m) \subset \mathrm{Pic}^0(A)$, $S(C)_a = \mathrm{Im}(a) \subset H^1(\mathcal{O}_A)$ and, with $\Pi_a : H^1_{DR}(A)_a \to H^1(\mathcal{O}_A)$ and $\Pi_m : H^1_{DR}(A)_m \to \mathrm{Pic}^0(A)$ as in (1.1) and (1.2), define

$$S_{DR}(C)_m = \Pi_m^{-1}(S(C)_m) \subset H^1_{DR}(A)_m$$
$$S_{DR}(C)_a = \Pi_a^{-1}(S(C)_a) \subset H^1_{DR}(A)_a$$

Both $S_{DR}(C)_m$ and $S_{DR}(C)_a$ contain $H^0(\Omega^1_{A/K})$ and by (1.6) we may consider the total space $S_{DR}(C)_t$ of the data consisting of $S_{DR}(C)_m$, $S_{DR}(C)_a$ and the inclusions $j_m : H^0(\Omega^1_{A/K}) \to S_{DR}(C)_m$, $j_a : H^0(\Omega^1_{A/K}) \to S_{DR}(C)_a$. Clearly $S_{DR}(C)_t$ is a K-linear subspace of the total de Rham space $H^1_{DR}(A)_t$ of (1.9).

One more definition: we say that C is a relative projective hull of A if the map m is injective and the map a is an isomorphism (the terminology is motivated by Serre's consideration of projective hulls of abelian varieties in: "Groupes Proalgebriques", Publ. Math. IHES, 7 (1960); our "relative projective hulls" are "truncations" of the projective hull of A, enjoying properties similar to those of a projective hull but relative to a given fixed subgroup of $\mathrm{Pic}^0(A)$. We will not go into details since they are irrelevant for our purposes). As we shall see in (2.5) below, $\mathcal{O}(C) = K$ if C is a relative projective hull of A.

Here is our main result:

(2.2) THEOREM. Fix a Lie K/k-algebra P and $D = K[P]$.

1) Let A be an abelian K-variety. Then the logarithmic Gauss-Manin $\mathrm{Der}_k K$-connection (1.2) is integrable. In particular the total de Rham space $H^1_{DR}(A)_t$ has an induced structure of D-module.

2) Let C be an irreducible commutative algebraic K-group with $\mathcal{O}(C) = K$ and abelian part A. Then C has at most one structure of algebraic D-group. It has a structure of algebraic

D-group if and only if $S_{DR}(C)_t$ is a D-submodule of $H^1_{DR}(A)_t$. In particular if C is a relative projective hull of A, then C has a unique structure of algebraic D-group.

3) Let C be an irreducible algebraic D-group with $\mathcal{O}(C) = K$. Then there is an isomorphism of D-modules

$$S_{DR}(C)_t \simeq L(C)^0$$

where $S_{DR}(C)_t$ is a D-module via the total Gauss-Manin connection, (cf. assertion 2) above) and $L(C)^0$ is a D-module via adjoint connection (cf. (I. 1.10)).

4) Let C_1, C_2 be irreducible algebraic D-groups with $\mathcal{O}(C_i) = K$, $i = 1,2$. Then any morphism of algebraic K-groups $C_1 \to C_2$ is automatically a morphism of algebraic D-groups. Moreover, if both C_i are relative projective hulls of A then we have a commutative diagram

$$\begin{array}{ccc} S_{DR}(C_2)_t & \simeq & L(C_2)^0 \\ \downarrow & & \downarrow \\ S_{DR}(C_1)_t & \simeq & L(C_1)^0 \end{array}$$

Consequently $H^1_{DR}(A)_t$ is isomorphic as a D-module with the continuous dual of

$$L(A^{hull}) := \varprojlim L(C)$$

where C runs through the set of relative projection hulls of A (note that these C's do not form a filtered projective system but their Lie algebras do !).

(2.3) REMARK. Recall that for any abelian K-variety A there exists a "universal extension" of A by a vector group called E(A) (see [MM]); E(A) is simply the extension of A by $G_a^{\dim A}$ defined by any linear isomorphism $X_a(G_a^{\dim A}) = K^{\dim A} \simeq H^1(\mathcal{O}_A)$. Clearly E(A) is a relative projective hull of A (in fact it is the "smallest" one i.e. a quotient of any relative projective hull). Note that $S_{DR}(E(A))_m = H^0(\Omega^1_{A/K})$ and $S_{DR}(E(A))_a = H^1_{DR}(A)_a = H^1_{DR}(A)$ so $S_{DR}(E(A))_t = H^1_{DR}(A)$. Then the above theorem says that

(∗) E(A) has a unique structure of algebraic D-group and we have an isomorphism of D-modules $H^1_{DR}(A) \simeq L(E(A))^0$ the left hand side being endowed with the usual Gauss-Manin connection, (1.1).

Now (∗) is easily seen to be a consequence of the crystalline theory developed in [MM] ("Grothendieck's Theorem") and of the duality theorem proved in the last chapter of [BBM]. But proving Theorem (2.2) along the linear of [MM] and [BBM] seems to us much harder.

In any case our proof of Theorem (2.2) (and hence of (∗)) is much less sophisticated: it is just a manipulation of Čech cocycles on a Zariski open covering of A (no crystalline background being necessary). It has the advantage of constructing explicitely the lifting of the operators $p \in P$ to E(A) and indeed to any relative projective hull of A ! In particular the D-module isomorphism $S_{DR}(C)_T \simeq L(C)^0$ from (2.2) should be viewed as a "dual" generalisation of the D-module counterpart of Grothendieck's theorem cited above. It also suggests that the following might hold:

(?) Any relative projective hull of an abelian variety has a crystalline nature.

(??) For any such relative projective hull C the isomorphism $S_{DR}(C)_t \simeq L(C)^0$ is "induced" be the crystalline structures.

Note also that our proof of (2.2) is mainly based on certain cohomological properties of A making very little use of the group law of A. This permits to pass from A to other varieties; this will be explained in a subsequent paper (and makes an essential diffeence between our approach and that in [MM]).

(2.4) We make a preparation. Assume C is as in (2.1) and write $B = B_m \times B_a$, B_m a torus of dimension M, B_a an algebraic vector group of dimension N. The projections $B \to B_m$ and $B \to B_a$ induce extensions of A by B_m and B_a respectively (call them C_m and C_a) which are described by the maps m = (m,0) and a = (0,a) respectively (in fact $C_m = C/B_a$ and $C_a = C/B_m$). If E' is a linear subspace of $E := X_a(B)$ then we have $E' = X_a(B')$ for a well defined quotient B' of B_a and the map $B_a \to B'$ induces an extension C' of A by B' ($C' = C_a/\mathrm{Ker}(B_a \to B')$) whose defining element $a' \in \mathrm{Hom}(E', H^1(\mathcal{O}_A))$ is the restriction of a to E'. Clearly, if $E = E' \oplus E''$, upon letting B", C", a" be the corresponding objects for E", the natural map

$$C \to C' \times_A C'' \times_A C_m$$

is an isomorphism. From now on we fix $E' = \mathrm{Ker}(a : E \to H^1(\mathcal{O}_A))$ and fix E" an arbitrary complement of E' in E. Since a' = 0 we have $C' = B' \times A$ so we may write

$$C = B' \times C^{\perp}$$

$$C^{\perp} = C'' \times_A C_m$$

It will be useful to recall how one obtains C via a glueing procedure starting from the maps (m,a). Start by lifting the map m to a homomorphism $(m_{ij})_{ij}$ from $\Sigma := X_m(B)$ to the Čech group $C^1(\mathcal{U}, \mathcal{O}_A^*) = \bigoplus_{i<j} H^0(U_i \cap U_j, \mathcal{O}_A^*)$ corresponding to a covering $\mathcal{U} = (U_i)_i$, $U_i = \mathrm{Spec}\, A_i$. Similarily lift a to a linear map $(a_{ij})_{ij}$ from $E = X_a(B)$ to $C^1(\mathcal{U}, \mathcal{O})$. Write $U_{ij} = U_i \cap U_j$ and put $A_{ij} = \mathcal{O}(U_{ij})$. Let R be any ring. Letting $R[\Sigma]$ be the group R-algebra on Σ and R[E] be the symmetric R-algebra of E and putting $R[\Sigma, E] = R[\Sigma] \otimes_R R[E]$ we have that C is obtained by glueing $\mathrm{Spec}(A_i[\Sigma, E])$ via the A_{ij} - isomorphisms $\phi_{ij} : A_{ij}[\Sigma, E] \to A_{ij}[\Sigma, E]$

(2.4.1) $\phi_{ij}(X) = m_{ij}(X)X$ for $X \in \Sigma$

(2.4.2) $\phi_{ij}(\xi) = \xi + a_{ij}(\xi)$ for all $\xi \in E$

Clearly C_m is described by glueing $\mathrm{Spec}(A_i[\Sigma])$ via (2.4.1) while C', C" are described by glueing $\mathrm{Spec}(A_i[E'])$, $\mathrm{Spec}(A_i[E''])$ via (2.4.2).

(2.5) LEMMA. In the above notations we have:

1) $\mathcal{O}(C) = K$ if and only if the maps m and a are injective.

2) $X_a(C^\perp) = 0$.

REMARK. Condition $\mathcal{O}(C) = K$ is equivalent to saying that C has no non-trivial affine quotient and also equivalent to saying that C has no non-trivial linear representation (cf. [Ro], [DG]).

Proof. Note first that any additive character $C^\perp \to G_a$ vanishes on B_m hence factors through $C^\perp/B_m = C''$ so 2) follows from 1); let's prove 1). Assume first that a is not injective; so in notations of (2.4) B' is non-trivial which implies $\mathcal{O}(C) \neq K$ in view of the decomposition $C = B' \times C^\perp$. Similarily, if m is not injective, put $\Sigma_1 = \mathrm{Ker}(m)$. Then letting $T_1 = \mathrm{Spec}(K[\Sigma_1])$ we have a surjective homomorphism $B \to T_1$ to which there corresponds a surjective homomorphism $C \to C_1$ where C_1 is a trivial extension of A by T_1. This implies again $\mathcal{O}(C) \neq K$. Conversely, assume m and a are injective and let's prove that $\mathcal{O}(C) = K$. We dispose of a commutative diagram with exact rows:

$$\begin{array}{ccccccccc} 0 \to & B_0 & \to & B & \to & B/B_0 & \to 0 \\ & \downarrow & & \downarrow & & \downarrow u & \\ 0 \to & C_0 & \to & C & \to & \mathrm{Spec}\,\mathcal{O}(C) & \to 0 \end{array}$$

where $\mathcal{O}(C_0) = K$ and B_0 is the linear part of C_0 (cf. [DG], p. 358). From the exact sequence

$$0 \to \mathrm{Ker}(u) \to C_0/B_0 \to C/B \to \mathrm{Coker}(u) \to 0$$

and from the fact that Ker (u) and Coker (u) are affine while $A_0 = C_0/B_0$ and $A = C/B$ are abelian varieties we get that $A_0 \to A$ is an isogeny. Now look at the commutative squares

$$\begin{array}{ccc} X_a(B) & \xrightarrow{f_a} & X_a(B_0) \\ a \downarrow & & \downarrow a_0 \\ H^1(\mathcal{O}_A) & \xrightarrow{\sim} & H^1(\mathcal{O}_{A_0}) \end{array} \qquad \begin{array}{ccc} X_m(B) & \xrightarrow{f_m} & X_m(B_0) \\ m \downarrow & & \downarrow m_0 \\ \mathrm{Pic}^0(A) & \xrightarrow[\text{isogeny}]{} & \mathrm{Pic}^0(A_0) \end{array}$$

Since a and m are injective while f_a and f_m are surjective it follows that f_a and f_m are isomorphisms hence $B_0 = B$ which implies $\mathrm{Spec}\,\mathcal{O}(C) = \mathrm{Spec}\,K$ and we are done

(2.6) LEMMA. Let G be an irreducible algebraic K-group with $\mathcal{O}(G) = K$. Then the map $\partial : P(G) \to \mathrm{Der}_k K$ is injective and $\mathrm{Der}_k G = L(G) \oplus P(G)$. In particular there is at most one structure of algebraic D-group on G.

Proof. By [Ro] G is commutative. By (I. 1.9) $\ker(\partial) = P(G/K) = X_a(G) \otimes L(G) = 0$. To prove that $\mathrm{Der}_k G = L(G) \oplus P(G)$, let $p \in \mathrm{Der}_k K$; since $\mathcal{O}(G) = K$ we have $pK \subset K$. Let m_e be the maximal ideal of $\mathcal{O}_{G,e}$ and let $\varepsilon : \mathcal{O}_{G,e} \to K = \mathcal{O}_{G,e}/m_e$ be the counit as usual. Then

$$\varepsilon p - p\varepsilon : \mathcal{O}_{G,e} \to K$$

is a K-ε-derivation hence it is of the form εv for a unique K-derivation $v : \mathcal{O}_G \to \mathcal{O}_G$, $v \in L(G)$. The equality

$$\varepsilon(p - v) = p\varepsilon$$

shows that $(p - v)(m_e) \subset m_e$. We claim that $d := p - v \in P(G)$. Indeed the map $\mu : \mathcal{O}_G \to \mu_* \mathcal{O}_{GxG}$ is an isomorphism (because $\mu = \Pi_2 \phi$ where $\phi : G \times G \to G \times G$, $\phi(x,y) = (x,xy)$ and $\Pi_2 : G \times G \to G$, $\Pi_2(x,y) = y$ and use Künneth formula to get $p_{2*}\mathcal{O}_{GxG} = \mathcal{O}_G$ which implies $\mu_* \mathcal{O}_{GxG} = \mathcal{O}_G$). Then

$$\mu d - (d \otimes 1 + 1 \otimes d)\mu \in \mathrm{Der}_K(\mathcal{O}_G, \mu_* \mathcal{O}_{GxG}) \simeq \mathrm{Der}_K(\mathcal{O}_G, \mathcal{O}_G) = \mathcal{O}(G) \otimes L(G) = L(G)$$

Since the above derivation (identified with a vector field in L(G)) vanishes at e (recall $d(m_e) \subset m_e$!) it must be zero so

$$\mu d = (d \otimes 1 + 1 \otimes d)\mu$$

Similarily one proves $Sd = dS$ consequently $d \in P(G)$ and we are done.

(2.7) LEMMA. Let G_1, G_2 be irreducible algebraic D-groups with $\mathcal{O}(G_i) = K$, $i = 1,2$. Then (with notations from (I. 2.1)):

$$\mathrm{Hom}_{\mathrm{alg}\, K\text{-gr}}(G_1^!, G_2^!) \simeq \mathrm{Hom}_{\mathrm{alg}\, D\text{-gr}}(G_1, G_2)$$

Proof. Let $f : G_1 \to G_2$ be a morphism of algebraic K-groups and $p \in P$. Then the morphism $pf - fp : \mathcal{O}_{G_2} \to f_* \mathcal{O}_{G_1}$ is a K-f-derivation (where we still denoted by p the derivation induced on $\mathcal{O}_{G_1}$ and $\mathcal{O}_{G_2}$ respectively) hence belongs to

$$\mathrm{Der}_K(\mathcal{O}_{G_2}, f_* \mathcal{O}_{G_1}) \simeq \mathrm{Hom}_K(\Omega^1_{G_2/K}, f_* \mathcal{O}_{G_1}) \simeq H^0(G_2, L(G_2) \otimes f_* \mathcal{O}_{G_1})$$

$$\simeq L(G_2) \otimes H^0(G_2, f_* \mathcal{O}_{G_1}) \simeq L(G_2) \otimes H^0(G_1, \mathcal{O}_{G_1}) \simeq L(G_2)$$

Moreover upon identifying $pf - fp$ with a vector field on G_2, it must vanish at $e \in G_2$ so it vanishes everywhere and we are done.

(2.8) In order to perform computations with cocycles, assume again we are in the situation of (2.4) so that we dispose of maps

$$(m_{ij}) : \Sigma = X_m(B) \to C^1(\mathcal{U}, \mathcal{O}_A^*)$$

$$(a_{ij}) : E = X_a(B) \to C^1(\mathcal{U}, \mathcal{O}_A)$$

lifting the maps m and a which define a given commutative algebraic K-group C. Moreover choose a basis $\chi_1, \dots, \chi_M$ of the $\mathbb{Z}$-module Σ, a K-basis $\xi_1 \dots, \xi_N$ of E and define $a^r_{ij} = a_{ij}(\xi_r)$ and $m^s_{ij} = m_{ij}(\chi_s)$. Upon refining $\mathcal{U}$ we may assume that the elements m_{ij} lift to some elements

$$(\omega^s_i, m^s_{ij}) \in \mathrm{Ker}(C^1_m(\mathcal{U}) \to C^2_m(\mathcal{U}))$$

notations being as in (1.2); indeed this is possible because $\Pi_m : H^1_{DR}(A)_m \to \mathrm{Pic}^0(A)$ is surjective (1.2). On the other hand a^r_{ij} automatically lift to elements

$$(\eta^r_i, a^r_{ij}) \in \mathrm{Ker}\,(C^1_a(\mathcal{U}) \to C^2_a(\mathcal{U}))$$

so recall that the following relations hold:

$$(2.8.1)_a \qquad da^r_{ij} = \eta^r_j - \eta^r_i$$

$$(2.8.1)_m \qquad (m^s_{ij})^{-1} dm^s_{ij} = \omega^s_j - \omega^s_i$$

For convenience, let's call a collection (a^r_{ij}, m^s_{ij}) as above a system of adapted cocycles for C. Then C is obtained by glueing the affine schemes

$$\mathrm{Spec}\, A_i[\xi_1, \dots, \xi_N, \chi_1, \chi_1^{-1}, \dots, \chi_M, \chi_M^{-1}] \quad (A_i = \mathcal{O}(U_i))$$

via the isomorphism ϕ_{ij}

$$(2.8.2)_a \qquad \phi_{ij}(\xi_r) = \xi_r + a^r_{ij}$$

$$(2.8.2)_m \qquad \phi_{ij}(\chi_s) = m^s_{ij}\chi_s$$

(2.9) LEMMA. Assume C is a relative projective hull of A and assume (a^r_{ij}, m^s_{ij}) is a system of adapted cocycles for C (2.8). Then for any $p \in \mathrm{Der}_k K$ there exist vector fields $v_k \in H^0(T_A)$, $1 \le k \le N$, liftings $\theta_i \in \mathrm{Der}_k A_i$ of p to $A_i = \mathcal{O}(U_i)$ and regular functions

$$\alpha^r_{jk}, \alpha^r_j, \mu^s_{jk}, \mu^s_j \in \mathcal{O}(U_i)$$

satisfying the following system of equations

$$(2.9.1) \qquad \sum_{r=1}^{N} a^r_{ij} v_r = \theta_j - \theta_i \qquad \text{for all } i,j$$

$$(2.9.2)_a \qquad v_k a^r_{ij} = \alpha^r_{jk} - \alpha^r_{ik} \qquad \text{for all } i,j \text{ and } 1 \le k,\, r \le n$$

$$(2.9.2)_m \qquad (m^s_{ij})^{-1} v_k m^s_{ij} = \mu^s_{jk} - \mu^s_{ik} \qquad \text{for all } i,j \text{ and } 1 \le k \le N,\ 1 \le s \le M$$

$$(2.9.3)_a \qquad \theta_j a^r_{ij} + \sum_{k=1}^{N} \alpha^r_{ik} a^k_{ij} = \alpha^r_j - \alpha^r_i \qquad \text{for all } i,j \text{ and } 1 \le r \le N$$

$$(2.9.3)_m \qquad (m^s_{ij})^{-1} \theta_j m^s_{ij} + \sum_{k=1}^{N} \mu^s_{ik} a^k_{ij} = \mu^s_j - \mu^s_i \qquad \text{for all } i,j \text{ and } 1 \le s \le M$$

Proof. Since the cup product

$$\cup : H^1(\mathcal{O}_A) \otimes H^0(T_A) \to H^1(T_A)$$

is an isomorphism one can find $v_1, \dots, v_n \in H^0(T_A)$ and liftings θ_i of p to derivations of $\mathcal{O}(U_i)$ such that (2.9.1) holds. By degeneration in E_1 of $E^{\cdot\cdot}(a)$, the map $d : H^1(\mathcal{O}_A) \to H^1(\Omega^1_{A/K})$ is zero. Hence the map $H^1(\mathcal{O}_A) \to H^1(\mathcal{O}_A)$ defined by applying v_k to cocycles is zero, in particular there exist $\tilde{\alpha}^r_{ik} \in \mathcal{O}(U_i)$ satisfying

$$v_k a^r_{ij} = \tilde{\alpha}^r_{jk} - \tilde{\alpha}^r_{ik} \quad \text{for all } i,j,k,r$$

Similarily, since the map dlog : $H^1(\mathcal{O}^*_A) \to H^1(\Omega^1_{A/K})$ sends the classes of (m^s_{ij}) into zero, one can find $\tilde{\mu}^s_{ik} \in \mathcal{O}(U_i)$ such that

$$(m^s_{ij})^{-1} v_k m^s_{ij} = \tilde{\mu}^s_{jk} - \tilde{\mu}^s_{ik} \qquad \text{for all } i,j,k,s$$

Now recall that we dispose of relations $(2.8.1)_a$ and $(2.8.1)_m$. To conclude the proof it would be sufficient to check that

$$(\theta_j a^r_{ij} + \sum_k \tilde{\alpha}^r_{ik} a^k_{ij})_{ij} \text{ and } ((m^s_{ij})^{-1}\theta_j m^s_{ij} + \sum_k \tilde{\mu}^s_{ik} a^k_{ij})_{ij}$$

are cocycles. Now we know from (1.1) and (1.2) that

$$((\theta_j - \theta_i) \quad \eta^r_i + \theta_i a^r_{ij})_{ij} \text{ and } ((\theta_j - \theta_i) \quad \omega^s_i + (m^s_{ij})^{-1}\theta_j m^s_{ij})_{ij}$$

are cocycles so what we have to check is that

$$((\theta_j - \theta_i) \wedge \eta^r_i - \sum \tilde{\alpha}^r_{ik} a^k_{ij})_{ij} \text{ and } ((\theta_j - \theta_i) \wedge \omega^s_i - \sum \tilde{\mu}^s_{ik} a^k_{ij})_{ij}$$

are cocycles.

Now taking contraction with v_k in $(2.8.1)_a$ we get

$$v_k \wedge \eta^r_j - v_k \wedge \eta^r_i = v_k a^r_{ij} = \tilde{\alpha}^r_{jk} - \tilde{\alpha}^r_{ik}$$

so $(v_k \wedge \eta^r_i - \tilde{\alpha}^r_{ik})_i$ glue together to give a constant $\lambda^r_k \in K$. Similarily $(2.8.1)_m$ implies that $\omega^s_i \wedge v_k - \tilde{\mu}^s_{ik} = c^s_k \in K$.

We get

$$(\theta_j - \theta_i) \wedge \eta^r_i - \sum \tilde{\alpha}^r_{ik} a^k_{ij} = \sum a^k_{ij} v_k \wedge \eta^r_i - \sum \tilde{\alpha}^r_{ik} a^k_{ij} = \sum_k \lambda^r_k a^k_{ij}$$

which is a cocycle and similarily

$$(\theta_j - \theta_i) \wedge \omega^s_i - \sum \tilde{\mu}_{ik} a^k_{ij} = \sum_k a^k_{ij} v_k \wedge \omega^s_i - \sum \tilde{\mu}_{ik} a^k_{ij} = \sum_k c^s_k a^k_{ij}$$

which is also a cocycle. Our lemma is proved.

(2.10) LEMMA. Assume C is any irreducible commutative algebraic K-group, (a^r_{ij}, m^s_{ij}) is a system of adapted cocycles for C and assume we have a solution

$$(v_k, \theta_i, \alpha^r_{jk}, \alpha^r_j, \mu^s_{jk}, \mu^s_j)$$

of the system (2.9.1)-(2.9.3). Define for each i a derivation $p_i \in \mathrm{Der}_k(A_i[E,\Sigma])$ by the formula

$$p_i = \theta_i + \sum_k \xi_k v_k - \sum_r \left(\sum_k \alpha^r_{ik}\xi_k + \alpha^r_i\right)\frac{\partial}{\partial \xi_r} - \sum_s \left(\sum_k \mu^s_{ik}\xi_k + \mu^s_i\right)\chi_s\frac{\partial}{\partial \chi_s}$$

Then the p_i's glue together to give a derivation $p_G \in \mathrm{Der}_k G$.

Proof. An easy computation using (2.8.2).

(2.11) COROLLARY. Let C be a relative projective hull of A. Then the map $\partial : P(C) \to \mathrm{Der}_k K$ is an isomorphism. In particular given any D, we have that C has a unique structure of algebraic D-group.

Proof. By (2.5) $\mathcal{O}(C) = K$ hence by (2.6) ∂ is injective. To check surjectivity of ∂ take $p \in \mathrm{Der}_k K$. By (2.9) and (2.10), p lifts to some derivation $p_G \in \mathrm{Der}_k G$. By (2.6) again $p_G = p^1_G + p^2_G$ with $p^1_G \in L(G)$, $p^2_G \in P(G)$; clearly p^2_G lifts p and we are done.

(2.12) LEMMA. Let C be an irreducible commutative algebraic K-group, $\mathcal{O}(C) = K$ and (a^r_{ij}, m^s_{ij}) be a system of adapted cocycles for C. Consider the K-linear space S of all systems $(\hat\theta, \hat\alpha^r_j, \hat\mu^s_j)_{1 \le r \le N, 1 \le s \le M}$ such that $\hat\theta \in L(A) = H^0(T_A)$ and $\hat\alpha^r_j$, $\hat\mu^s_j \in \mathcal{O}(U_i)$ satisfy the following system:

$$(2.12)_a \qquad \hat\theta a^r_{ij} = \hat\alpha^r_j - \hat\alpha^r_i, \qquad 1 \le r \le N$$

$$(2.12)_m \qquad (m^s_{ij})^{-1}\hat\theta m^s_{ij} = \hat\mu^s_j - \hat\mu^s_i, \qquad 1 \le s \le M$$

Moreover, for any such $(\hat\theta, \hat\alpha^r_j, \hat\mu^s_j) \in S$ consider the derivations $\hat p_j \in \mathrm{Der}(A_j[E,\Sigma])$ defined by the formula

$$\hat p_j = \hat\theta - \sum_r \hat\alpha^r_j \frac{\partial}{\partial \xi_r} - \sum_s \hat\mu^s_j \chi_s \frac{\partial}{\partial \chi_s}$$

Then:

1) The $\hat p_j$'s glue together to give a derivation $\hat p \in L(C)$ (we write $\hat p = \{\hat\theta, \hat\alpha^r_j, \hat\mu^s_j\}$);

2) The map assigning to each element of S the corresponding derivation in L(C) is a K-linear isomorphism.

Proof. 1) is again an immediate computation using (2.8.2).

2) follows by noting that the map $S \to L(C)$ is injective and that $\dim_K S = \dim_K L(A) + N + M = \dim C = \dim_K L(C)$.

(2.13) To formulate the next Lemma let's make the following definition. Assume P is a Lie K/k-algebra and V, W are two finite dimensional K-linear spaces with P-connections. By a

P-duality between V and W we will mean a bilinear non-degenerate form

$$\langle\, ,\rangle : V \times W \to K$$

such that

$$p\langle x,y\rangle = \langle px,y\rangle + \langle x,py\rangle$$

for all $p \in P$, $x \in V$, $y \in W$. We will speak about orthogonals in the usual sense. Note that a K-linear subspace V_1 of V is stable under P if and only if its orthogonal $V_1^{\perp} \subset W$ is stable under P. Moreover the connection on V is integrable if and only if the connection on W is so.

Here is our main step in proving Theorem (2.2):

(2.14) LEMMA. Assume C is a relative projective hull of A, P is any Lie K/k-algebra and $D = K[P]$. Then there is a P-duality

$$\langle\, ,\rangle : S_{DR}(C)_t \times L(C) \to K$$

where $S_{DR}(C)_t$ has a P-connection induced from that of $H^1_{DR}(A)_t$ (cf. (1.9)) and L(C) has a P-connection defined by the adjoint connection (cf. (I.1.10) where C is viewed with its unique structure of algebraic D-group, cf. (2.11)). Moreover, for any unipotent linear algebraic subgroup B_1 of C we have

$$L(B_1)^{\perp} = S_{DR}(C/B_1)_t$$

hence we have an induced P-duality

$$\langle\, ,\rangle : S_{DR}(C/B_1)_t \times L(C/B_1) \to K$$

Proof. Consider a system of adapted cocycles (a^r_{ij}, m^s_{ij}) for C as in (2.8) from where we borrow notations. We also use notations from (1.1), (1.2). Let $\tilde{Z}^1_m(\mathcal{U})$ be the subgroup of $Z^1_m(\mathcal{U}) :=$ $= \operatorname{Ker}(C^1_m(\mathcal{U}) \to C^2_m(\mathcal{U}))$ of those elements $(\omega_i, y_{ij}) \in C^1_m(\mathcal{U})$ such that

$$(2.14.1)_m \qquad y_{ij} = \left(\prod_s (m^s_{ij})^{n_s}\right) z_j z_i^{-1}$$

for some integers n_s and some $z_i \in \mathcal{O}^*(U_i)$. Clearly the map $\tilde{Z}^1_m(\mathcal{U}) \to S_{DR}(C)_m$ is surjective (due to our assumption that our cocycles are "adapted"). On the other its kernel is easily seen to coincide with $B^1_m(\mathcal{U}) = \operatorname{Im}(C^0_m(\mathcal{U}) \to C^1_m(\mathcal{U}))$ so after all we have

$$S_{DR}(C)_m = \tilde{Z}^1_m(\mathcal{U})/B^1_m(\mathcal{U})$$

Of course since the map $a : X_a(B) \to H^1(\mathcal{O}_A)$ was assumed to be an isomorphism we have

$$S_{DR}(C)_a = H^1_{DR}(A)_a = Z^1_a(\mathcal{U})/B^1_a(\mathcal{U})$$

where $Z^1_a(\mathcal{U}) = \operatorname{Ker}(C^1_a(\mathcal{U}) \to C^2_a(\mathcal{U}))$, $B^1_a(\mathcal{U}) = \operatorname{Im}(C^0_a(\mathcal{U}) \to C^1_a(\mathcal{U}))$. We will define a $\mathbf{Z}$ - bilinear map

$$\langle\, ,\,\rangle : (\tilde{Z}^1_m(\mathcal{U}) \oplus Z^1_a(\mathcal{U})) \times L(C) \to K$$

Let $(\omega_i, y_{ij}) \in \tilde{Z}^1_m(\mathcal{U})$, $(\eta_i, x_{ij}) \in Z^1_a(\mathcal{U})$ and denote by $\{\hat{\theta}, \hat{\alpha}^r_j, \hat{\mu}^s_j\} \in L(C)$ a derivation defined as in (2.12). We have

$$(2.14.1)_a \qquad x_{ij} = \sum_k \lambda_k a^k_{ij} + u_j - u_i$$

for some $\lambda_k \in K$ and $u_j \in \mathcal{O}(U_j)$. Then define (using notations from $(2.14.1)_a$ and $(2.14.1)_m$):

$$(2.14.2) \qquad \langle(\omega_j, y_{ij}, \eta_j, x_{ij}), \{\hat{\theta}, \hat{\alpha}^r_j, \hat{\mu}^s_j\}\rangle =$$

$$= \hat{\theta} \wedge \omega_j - \sum_s n_s \hat{\mu}^s_j - z_j^{-1} \hat{\theta} z_j + \hat{\theta} \wedge \eta_j - \sum_k \lambda_k \hat{\alpha}^k_j - \hat{\theta} u_j$$

Of course we must prove that the right hand side of the above equality is independent of j and hence belongs to K. But indeed

$$\hat{\theta} \wedge \omega_j - \hat{\theta} \wedge \omega_i = \hat{\theta} \wedge (\omega_j - \omega_i) = \hat{\theta} \wedge (y_{ij}^{-1} dy_{ij}) =$$

$$= y_{ij}^{-1} \hat{\theta} y_{ij} = \sum n_s (m^s_{ij})^{-1} \hat{\theta} m_{ij} + z_j^{-1} \hat{\theta} z_j - z_i^{-1} \hat{\theta} z_i$$

Consequently

$$\hat{\theta} \wedge \omega_j - \hat{\theta} \wedge \omega_i = \sum n_s (\hat{\mu}^s_j - \hat{\mu}^s_i) + z_j^{-1} \hat{\theta} z_j - z_i^{-1} \hat{\theta} z_i \,.$$

Similarily we get

$$\hat{\theta} \wedge \eta_j - \hat{\theta} \wedge \eta_i = \sum \lambda_k (\hat{\alpha}^k_j - \hat{\alpha}^k_j) + \hat{\theta} u_j - \hat{\theta} u_i \,.$$

Summing up the last two equalities we get that the right hand side member of (2.14.2) is independent of j as required. It is easy to check that our $\mathbf{Z}$ - bilinear map induces a non-degenerate K-bilinear map

$$\langle\, ,\,\rangle : S_{DR}(C)_t \times L(C) \to K$$

It is also easy to check that $L(B_1)^\perp = S_{DR}(C/B_1)_t$ for any unipotent $B_1 \subset C$: first one shows that our definition of $\langle\, ,\,\rangle$ does not depend on choosing the basis $\xi_1, \ldots, \xi_N$ of $X_a(B)$ in (2.8); next we choose a basis with the property that $X_a(B/B_1)$ admits $\xi_1, \ldots, \xi_{N_1}$ $(N_1 \le N)$ as a basis. Since dimensions of $L(B_1)^\perp$ and $S_{DR}(C/B_1)_t$ are equal it is sufficient to check that

$$\langle S_{DR}(C/B_1)_t, L(B_1)\rangle = 0$$

which follows from definitions using our special basis.

It remains to prove that $\langle\, ,\,\rangle$ is a P-duality. This can be done by a tedious (but not straightforward) computation with cocycles; we give the outline of this computation since we felt it would have been unfair to just "leave it to the reader".

Pick any $p \in P$, choose a solution

$$(v_k,\theta_j,\alpha^r_{jk},\alpha^r_j,\mu^s_{jk},\mu^s_j)$$

of the system (2.9.1)-(2.9.3) with the θ_j's lifting $\partial(p) \in \mathrm{Der}_k K$ and pick closed 1-forms η^k_j and ω^s_j satisfying

(2.14.3) $$da^k_{ij} = \eta^k_j - \eta^k_i, \quad (m^s_{ij})^{-1} dm^s_{ij} = \omega^s_j - \omega^s_i$$

Then what we must prove is that the expression

(2.14.4) $$\partial(p) < \{\omega^s_j, m^s_{ij}, \eta^r_j, a^r_{ij}\}, \{\hat\theta, \hat\alpha^r_j, \hat\mu^s_j\} >$$

equals the sum of (2.14.5) and (2.14.6) below:

(2.14.5) $$< \nabla_p \{\omega, m, \eta, a\}, \{\hat\theta, \hat\alpha, \hat\mu\} >$$

(2.14.6) $$< \{\omega, m, \eta, a\}, \nabla_p \{\hat\theta, \hat\alpha, \hat\mu\} >$$

where { } was also used to denote classes in $S_{DR}(C)_t$ and we made the obvious abreviations $\omega = (\omega^s_j), \ldots$.

Now (2.14.4) equals

(2.14.7) $$\theta_j(\hat\theta \wedge \omega^s_j - \hat\mu^s_i + \hat\theta \wedge \eta^r_j - \hat\alpha^r_j)$$

Next (2.14.5) becomes

(2.14.8) $$<\{0, 0, \mathrm{Lie}_{\theta_j} \omega^s_j + \mathrm{Lie}_{\theta_j} \eta^r_j, (\theta_j - \theta_i) \wedge (\omega^s_j + \eta^r_j) + \theta_i a^r_{ij} + (m^s_{ij})^{-1} \theta_i m^s_{ij}\}, \{\hat\theta, \hat\alpha, \hat\mu\} >$$

To compute (2.14.6) note first that since $\mathrm{Der}_k G = L(G) \oplus P(G)$, cf. (2.6) we have

$$\nabla_p \{\hat\theta, \hat\alpha, \hat\mu\} = [p_G, \hat p] \qquad (\text{Lie bracket in } \mathrm{Der}_k C)$$

where p_G is given by the formula in (2.10) and $\hat p$ is defined by the formula in (2.12). A long computation yields:

$$[p_G, \hat p] = \{[\theta_j, \hat\theta] + \sum_k \hat\alpha^k_j v_k, \sum_k \alpha^r_{jk} \hat\alpha^k_j + \theta_j \hat\alpha^r_j - \hat\theta \alpha^r_j, \sum_k \mu^s_{jk} \hat\alpha^k_j + \theta_j \hat\mu^s_j - \hat\theta \mu^s_j\}$$

so (2.14.6) becomes

(2.14.9) $$([\theta_j, \hat\theta] + \sum_k \hat\alpha^k_j v_k) \wedge (\omega^s_j + \eta^r_j) - \sum_k \alpha^r_{jk} \hat\alpha^k_j - \theta_j \hat\alpha^r_j + \hat\theta \alpha^r_j - \sum_k \mu^s_{jk} \hat\alpha^k_j - \theta_j \hat\mu^s_j + \hat\theta \mu^s_j$$

Finally let $\lambda_k \in K$ and $u_j \in \mathcal{O}(U_j)$ such that

(2.14.10) $$(\theta_j - \theta_i) \wedge (\omega^s_j + \eta^r_j) + \theta_i a^r_{ij} + (m^s_{ij})^{-1} \theta_i m^s_{ij} = \sum \lambda_k a^k_{ij} + u_j - u_i$$

Then by the very definition (2.14.8) becomes

(2.14.11) $$\hat\theta \wedge (\mathrm{Lie}_{\theta_j}(\omega^s_j + \eta^r_j)) - \sum_k \lambda_k \hat\alpha^k_j - \hat\theta u_j$$

By well known properties of Lie derivative we have

$$\theta_j(\hat{\theta}\wedge\phi) = [\theta_j,\hat{\theta}]\wedge\phi + \hat{\theta}\wedge(\mathrm{Lie}_{\theta_j}\phi)$$

for any 1-form ϕ. In view of this, in order to prove that (2.14.7) equals (2.14.9) + (2.14.11) it is sufficient to check the following (with $\phi = \omega_j^s + \eta_j^r$):

$$\sum_k (v_k\wedge\phi - \alpha_{jk}^r - \mu_{jk}^s - \lambda_k)\hat{\alpha}_j^k + \hat{\theta}(\alpha_j^r + \mu_j^s - u_j) = 0. \tag{2.14.12}$$

Now introducing (2.9.1), $(2.9.3)_a$, $(2.9.3)_m$ in (2.14.10) we get

$$\sum_k (v_k\wedge\phi - \alpha_{jk}^r - \mu_{jk}^s - \lambda_k)a_{ij}^k = (u_j - \alpha_j^r - \mu_j^s) - (u_i - \alpha_i^r - \mu_i^s) \tag{2.14.13}$$

Exactly as in the proof of (2.9) we get that

$$v_k\wedge\omega_j^s - \mu_{jk}^s \in K$$

$$v_k\wedge\eta_j^r - \alpha_{jk}^r \in K$$

hence we get (using the fact that the cohomology classes of a_{ij}^k form a basis in $H^1(\mathcal{O}_A)$) that:

$$v_k\wedge\phi - \alpha_{jk}^r - \mu_{jk}^s - \lambda_k = 0$$

$$u_j - \alpha_j^r - \mu_j^s = u_i - \alpha_i^r - \mu_i^s \in K$$

which clearly implies (2.14.12) and our lemma is proved.

(2.15) Let's pass to the proof of the duality theorem (2.2). By (2.14) for any relative projective hull C of A we have an isomorphism of K-linear spaces with P-connection $S_{DR}(C)_t \simeq L(C)^o$. Assume now that C_1 and C_2 are two relative projective hulls of A. We will say that C_1 is bigger than C_2 if $S(C_1)_m \otimes Q$ contains $S(C_2)_m \otimes Q$; if this is the case then there is a natural Lie algebra map $L(C_1) \to L(C_2)$ defined as follows. Choose any integer $n \geq 1$ such that $nS(C_2)_m \subset S(C_1)_m$ (as subgroups $\mathrm{Pic}^o(A)$) and let C_3 be the relative projective hull of A for which $S(C_3)_m = nS(C_2)_m$. Then there is an isogeny $C_2 \to C_3$ and a morphism $C_1 \to C_3$ so there is a Lie algebra map $L(C_1) \to L(C_3) \simeq L(C_2)$; one can easily check that this map does not depend on the choice of the integer n. Moreover one checks that the induced square of spaces with P-connection

$$\begin{array}{ccc} S_{DR}(C_1)_t & \simeq & L(C_1)^o \\ \uparrow & & \uparrow \\ S_{DR}(C_2)_t & \simeq & L(C_2)^o \end{array}$$

is commutative. Passing to direct limit we get

$$H^1_{DR}(A)_t = \varinjlim_C S_{DR}(C)_t = \varinjlim_C L(C)^o = L(A^{\mathrm{hull}})$$

Since the P-connection on $L(A^{hull})$ is obviously integrable so will be the P-connection on $H^1_{DR}(A)_t$ which proves assertion 1) in (2.2). Next let C be any irreducible commutative algebraic K-group with abelian part A and with $\mathcal{O}(C) = K$. Then by (2.5) the maps $m : X_m(B) \to Pic^0(A)$ and $X_a(B) \to H^1(\mathcal{O}_A)$ (B = linear part of C) are injective so if we let C_1 be the relative projective hull of A for which $S(C_1)_m = S(C)_m$ there will be a surjective morphism $C_1 \to C$ whose kernel B_1 is unipotent. Consider the P-duality from (2.14)

$$\langle\, ,\rangle : S_{DR}(C_1)_t \times L(C_1) \to K$$

By (2.14) $S_{DR}(C)_t = S_{DR}(C_1/B_1)_t = L(B_1)^{\perp}$ so $S_{DR}(C)_t$ is a D-submodule of $S_{DR}(C_1)_t$ (equivalently of $H^1_{DR}(A)_t$) if and only if $L(B_1)$ is a D-submodule of $L(C_1)$. By (I.1.16) the latter happens if and only if B_1 is an algebraic D-subgroup of C_1 which by (I.1.17) happens if and only if C has an algebraic D-group structure such that $C_1 \to C$ is a D-map. By (2.7) and (2.11) the latter condition is equivalent to C having an algebraic D-group structure. This together with (2.6) and (2.11) proves assertion 2) in Theorem (2.2) as well as the assertion 3). Assertion 4) follows from (2.7) and from the above discussion involving $L(A^{hull})$. Our Theorem (2.2) is proved.

Now we pass to our applications.

(2.16) For any irreducible commutative algebraic K-group C we let $Der_k^{S(C)}K$ be the Lie K/k-subalgebra of $Der_k K$ consisting of all derivations $p \in Der_k K$ for which

$$\ell\nabla_p(S_{DR}(C)_m) \subset S_{DR}(C)_a$$

$$\nabla_p(S_{DR}(C)_a) \subset S_{DR}(C)_a$$

where $(\ell\nabla, \nabla)$ is the logarithmic $Der_k K$-connection on $(H^1_{DR}(A)_m, H^1_{DR}(A)_a)$ cf. (1.2). Equivalently $Der_k^{S(C)}K$ is the Lie K/k-subalgebra of $P = Der_k K$ consisting of all $p \in P$ which send $S_{DR}(C)_t$ into itself (where we view $S_{DR}(C)_t$ as a subspace of the D-module $H^1_{DR}(A)_t$, $D = K[P]$).

(2.17) COROLLARY. Let C be an irreducible commutative algebraic K-group. Then there is a split exact sequence of Lie K/k-algebras

$$0 \to X_a(C) \otimes L(C) \to P(C) \overset{\partial}{\to} Der_k^{S(C)}K \to 0$$

Proof. Let as in (2.4) $E' = Ker(a : E \to H^1(\mathcal{O}_A))$ and $E'' \subset E$ be a complement of E' in E. Moreover, let $\Sigma' = Ker(m : \Sigma \to Pic^0(A))$, $S_m = Im(m : \Sigma \to Pic^0(A))$, choose a basis of the $\mathbf{Z}$-module $S_m/Tors(S_m)$, lift it to a subset of Σ and Σ'' be the subgroup generated in Σ by this subset; hence $\Sigma' \cap \Sigma'' = 0$, $[\Sigma : \Sigma' \oplus \Sigma''] < \infty$ and the restriction of m to Σ'' is injective. The inclusion $\Sigma' \oplus \Sigma'' \subset \Sigma$ induces an isogeny $B_m = Spec\,K[\Sigma] \to B'_m \times B''_m$ where $B'_m = Spec\,K[\Sigma']$. Hence we get an isogeny:

$$B = B_m \times B_a \to B'_m \times B''_m \times B'_a \times B''_a$$

where $B'_a = Spec\,K[E']$, $B''_a = Spec\,K[E'']$. To this isogeny there corresponds an isogeny $C \to C_1$

over A where $C_1 = B'_a \times B'_m \times C_2$ and the maps m_2 and a_2 corresponding to C_2 are injective. By (2.5) $\mathcal{O}(C_2) = K$. Clearly $S_{DR}(C_1)_t = S_{DR}(C_2)_t$. By (I.1.19) we have $\mathrm{Im}(P(C_1) \to \mathrm{Der}_k K) = \mathrm{Im}(P(C_2) \to \mathrm{Der}_k K)$ and there is a Lie K/k-algebra map from the latter space to $P(C_1)$ (use the fact that $B'_m \times B'_a$ is defined over Q!). By (2.2) $P(C_2) \simeq \mathrm{Der}_k^{S(C_2)} K$ so we get a split exact sequence of Lie K/k-algebras

$$0 \to P(C_1/K) \to P(C_1) \overset{\partial}{\to} \mathrm{Der}_k^{S(C_1)} K \to 0$$

Finally note that $S_{DR}(C)_t = S_{DR}(C_1)_t$ and we conclude by (I.1.21).

(2.18) COROLLARY. Let C be an irreducible commutative algebraic K-group, P a Lie K/k-algebra and D = K[P]. Then C has a structure of algebraic D-group if and only if

$$\partial(p) \in \mathrm{Der}_k^{S(C)} K \qquad \text{for all } p \in P.$$

If $\mathcal{O}(C) = K$, this structure is unique and is given by the formula in (2.10).

(2.19) REMARK. One can give an analytic interpretation of the unique structure of algebraic D-group on the universal extension E(A) of an abelian variety A (cf. (2.3)) in case $k = \mathbf{C}$ and $A = \mathcal{A} \times_X \mathrm{Spec}\, K$ where $\mathcal{A} \to X$ is an abelian scheme and K is the algebraic closure of k(X). Indeed, in this case (upon replacing X by an étale open set of it) we get that $E(A) = \tilde{\mathcal{A}} \times_X \mathrm{Spec}\, K$ where $\tilde{\mathcal{A}} \to X$ is analytically (but not algebraically) isomorphic over X^{an} with an analytically trivial bundle with fibre $(\mathbf{C}^*)^{2\dim(A)}$. Then any analytic vector field v on X lifts to an analytic vector field $\tilde{v}$ on $(\tilde{\mathcal{A}})^{an}$ which agrees with multiplication and inverse maps. The remarkable fact is that if v is algebraic then $\tilde{v}$ is also algebraic; this is of course a corollary of (2.2) and indeed of Groethendieck's theorem [MM]. Our method has the advantage of giving a formula for $\tilde{v}$ in terms of v cf. (2.9) and (2.10).

(2.20) We close by illustrating (2.18) in the simplest case.

EXAMPLE

Let A be an elliptic curve over K, with j-invariant $j \in K$, let $C = E(A)$ be its universal extension and let $p \in \mathrm{Der}_k K$, $pj \neq 0$. We will indicate what is the explicit lifting of p to P(C), cf. (2.18), (2.10). Cover A by two affine open sets U_1, U_2, choose any liftings $\theta_1 \in \mathrm{Der}_k \mathcal{O}(U_1)$, $\theta_2 \in \mathrm{Der}_k \mathcal{O}(U_2)$ of p and choose a global vector field θ on A. Then by (2.10) there exist $a \in \mathcal{O}(U_1 \cap U_2)$, $\alpha_1, \beta_1 \in \mathcal{O}(U_1)$, $\alpha_2, \beta_2 \in \mathcal{O}(U_2)$ such that the following hold:

$$(*) \qquad \begin{aligned} a\theta &= \theta_2 - \theta_1 \\ \theta a &= \alpha_2 - \alpha_1 \\ \theta_2 a + \alpha_1 a &= \beta_2 - \beta_1 \end{aligned}$$

Define:

$$p_1 = \theta_1 + \xi\theta - (\alpha_1\xi + \beta_1)\frac{\partial}{\partial\xi} \in \mathrm{Der}_k\mathcal{O}(U_1)[\xi]$$

$$p_2 = \theta_2 + \xi\theta - (\alpha_2\xi + \beta_2)\frac{\partial}{\partial\xi} \in \mathrm{Der}_k\mathcal{O}(U_2)[\xi]$$

and view C as obtained by glueing $U_1 \times \mathbf{A}^1$ and $U_2 \times \mathbf{A}^1$, $\mathbf{A}^1 = \mathrm{Spec}\, K[\xi]$ via the $\mathcal{O}(U_1 \cap U_2)$-algebra map

$$\phi : \mathcal{O}(U_1 \cap U_2)[\xi] \to \mathcal{O}(U_1 \cap U_2)[\xi]$$

$$\phi(\xi) = \xi + a$$

Then p_1 and p_2 glue together to give our lifting p_C of p to P(C).

So the problem of computing "explicitely" p_C for a given A and p amounts to finding a, α_1, β_1, α_2, β_2 satisfying the system (*) above. If A is the smooth projective model of $U_1 = \mathrm{Spec}\, K[x,y]/(y^2 - x(x-1)(x-\lambda))$ then this computation can be done explicitely. Rather than performing this computation (whose final result is after all irrelevant) we shall indicate the steps which should be followed and then leave it to the reader who enjoys contemplating formulae. A can be covered by U_1 and $U_2 = \mathrm{Spec}\, K[u,v]/(u - v(v-u)(v-\lambda u))$ glued via the formulae

$$u = \frac{1}{y}, \quad v = \frac{x}{y}$$

We may take of course $\theta = y\frac{\partial}{\partial x}$; solving the system (*) reduces to two problems:

a) lifting explicitely p to $\mathrm{Der}_k\mathcal{O}(U_1)$, $\mathrm{Der}_k\mathcal{O}(U_2)$.

b) finding (in an explicit way) for any $\phi_1, \phi_2 \in \mathcal{O}(U_1 \cap U_2)$ a relation of the form $\lambda_1\phi_1 + \lambda_2\phi_2 = a_2 - a_1$ where $(\lambda_1, \lambda_2) \in K^2 \setminus \{(0,0)\}$, $a_1 \in \mathcal{O}(U_1)$, $a_2 \in \mathcal{O}(U_2)$.

Indeed by a) we find θ_1, θ_2 explicitely and put $a = (\theta_2 - \theta_1)/y\frac{\partial}{\partial x}$.

Then by b) we write explicitely $y\frac{\partial a}{\partial x} = \tilde{\alpha}_2 - \tilde{\alpha}_1$, where $\tilde{\alpha}_1 \in \mathcal{O}(U_1)$, $\tilde{\alpha}_2 \in \mathcal{O}(U_2)$. Finally by b) we write explicitely $\theta_2 a + \tilde{\alpha}_1 a + \lambda a = \beta_2 - \beta_1$ for some $\lambda \in K$, $\beta_1 \in \mathcal{O}(U_1)$, $\beta_2 \in \mathcal{O}(U_2)$ and define $\alpha_1 = \tilde{\alpha}_1 + \lambda$, $\alpha_2 = \tilde{\alpha}_2 + \lambda$.

Let's deal separately with problems a) and b). For problem a) let p^* denote the lifting of p to K[x,y] which kills x,y; then one can determine, explicitely by a standard procedure, polynomials $M,N,Q \in K[x,y]$ such that

$$Mf + N\frac{\partial f}{\partial x} + Q\frac{\partial f}{\partial y} = 1$$

where $f = y^2 - x(x-1)(x-\lambda)$ and note that the derivation

$$p^* - p^*(f)N\frac{\partial}{\partial x} - p^*(f)Q\frac{\partial}{\partial y} \in \mathrm{Der}_k K[x,y]$$

sends the ideal (f) into itself. This derivation induces a lifting θ_1 of p to $\mathrm{Der}_k\mathcal{O}(U_1)$; θ_2 is constructed analogously.

To deal with problem b) note that (upon denoting by $\bar{x}$, $\bar{y}$ the classes of x, y modulo (f)) we have

$$\mathcal{O}(U_1 \cap U_2) = K[\bar{x},\bar{y},\bar{y}^{-1}] = (\sum_{n\in\mathbf{Z}} K\bar{y}^n) + (\sum_{n\in\mathbf{Z}} K\bar{x}\bar{y}^n) + (\sum_{n\in\mathbf{Z}} K\bar{x}^2\bar{y}^n)$$

Now note that

$$\bar{y}^n \in K[\bar{x},\bar{y}] = \mathcal{O}(U_1) \qquad \text{for } n \geq 0$$

$$\bar{y}^n \in K[\bar{x}/\bar{y},1/\bar{y}] = \mathcal{O}(U_2) \qquad \text{for } n < 0$$

$$\bar{x}\bar{y}^n \in \mathcal{O}(U_1) \qquad \text{for } n \geq 0$$

$$\bar{x}\bar{y}^n \in \mathcal{O}(U_2) \qquad \text{for } n < 0$$

$$\bar{x}^2\bar{y}^n \in \mathcal{O}(U_1) \qquad \text{for } n \geq 0$$

$$\bar{x}^2\bar{y}^n \in \mathcal{O}(U_2) \qquad \text{for } n < -1$$

Consequently any element of $K[\bar{x},\bar{y},\bar{y}^{-1}]$ can be explicitely written as a sum

$$\lambda\bar{x}^2/\bar{y} + a_2 - a_1$$

where $\lambda \in K$, $a_1 \in K[\bar{x},\bar{y}]$, $a_2 \in K[\bar{x}/\bar{y},1/\bar{y}]$ and we are done (note that $\bar{x}^2/\bar{y}$ appears as a "canonical" Čech representative for $H^1(\mathcal{O}_A)$!).

Similar computations can be performed providing the explicit lifting of a derivation $p \in \mathrm{Der}_k K$ to $P(C)$ where C is the extension of A (= projective smooth model of $y^2 - x(x-1)(x-\lambda) = 0$, $\lambda \in K$, $p\lambda \neq 0$) by $G_m \times G_a$ with $\mathcal{O}(C) = K$ corresponding to a given line bundle $\mathcal{O}(Q - Q_o)$ on A ($Q,Q_o \in A(K)$).

3. Descent. Regularity

We continue with our applications of the preceeding theory. Start with a preparation.

(3.1) Let V be a smooth projective K-variety and K_o an algebraically closed field of definition for V. Then there is an integrable logarithmic $\mathrm{Der}_{K_o} K$-connection on $(H^1(\mathcal{O}_V^*), H^1(\mathcal{O}_V))$ defined as follows. Fix a descent isomorphism $V \simeq V_o \otimes_{K_o} K$, V_o a smooth projective K_o-variety, view V as a split $K[P_o]$-variety, $P_o = \mathrm{Der}_{K_o} K$ and for $p \in P_o$ define maps:

$$\nabla_p : H^1(\mathcal{O}_V) \to H^1(\mathcal{O}_V)$$

$$\ell\nabla_p : H^1(\mathcal{O}_V^*) \to H^1(\mathcal{O}_V)$$

by the formulae

$$\nabla_p([a_{ij}]) = [pa_{ij}]$$

$$\ell\nabla_p([m_{ij}]) = [m_{ij}^{-1} p m_{ij}]$$

where $[x_{ij}]$ denotes Čech cohomology class of the cocycle (x_{ij}). The link between the above logarithmic connection and the logarithmic Gauss-Manin connection is given by the following:

(3.2) LEMMA. With notations from (1.2) and (3.1), for any $p \in \mathrm{Der}_{K_o} K$ we have the following commutative diagrams:

$$\begin{array}{ccc} H^1_{DR}(V)_a & \xrightarrow{\nabla_p} & H^1_{DR}(V)_a \\ \Pi_a \downarrow & & \downarrow \Pi_a \\ H^1(\mathcal{O}_V) & \xrightarrow[\nabla_p]{} & H^1(\mathcal{O}_V) \end{array}$$

$$\begin{array}{ccc} H^1_{DR}(V)_m & \xrightarrow{\ell\nabla_p} & H^1_{DR}(V)_a \\ \Pi_m \downarrow & & \downarrow \Pi_a \\ \mathrm{Pic}^o(V) \subset H^1(\mathcal{O}_V^*) & \xrightarrow{\ell\nabla_p} & H^1(\mathcal{O}_V) \end{array}$$

Proof. Use only definitions.

The above lemma shows in particular that the restriction of $(\nabla, \ell\nabla)$ to a logarithmic $\mathrm{Der}_{K_o} K$-connection on $(\mathrm{Pic}^o(V), H^1(\mathcal{O}_V))$ does not depend on the choice of the "descent isomorphism" $V \simeq V_o \otimes_{K_o} K$ we have fixed. The following lemma provides another useful description of our map $\ell\nabla$ defined in (3.1):

(3.3) LEMMA. With notations from (3.1) identify $H^1(\mathcal{O}_V^*)$ with the group of K-points of the (locally algebraic) K_o-group scheme $A = \mathrm{Pic}_{V_o/K_o}$ and identify $H^1(\mathcal{O}_V)$ with the group of K-points of the Lie algebra $L(A)$. Then the map $\ell\nabla_p : H^1(\mathcal{O}_V^*) \to H^1(\mathcal{O}_V)$ defined for each $p \in \mathrm{Der}_{K_o} K$ in (3.1), restricted to $A^o(K)$ ($A^o = \mathrm{Pic}^o_{V_o/K_o}$) identifies with Kolchin's logarithmic derivative $\ell\nabla_{p^*} : A^o(K) \to L(A^o)$, see (I.1.12).

Proof. To check the above statement it is useful to adopt a functorial viewpoint. So let

$$\underline{A} : \{\text{locally noetherian } K_o\text{-schemes}\} \to \{\text{groups}\}$$

be a contraviarint functor. Then one can define for any $p \in \mathrm{Der}_{K_o} K$ the "logarithmic derivative"

$$\ell_{\underline{A}} p : \underline{A}(K) \to L_K(\underline{A})$$

(where $L_K(\underline{A})$ is the kernel of the map $\underline{A}(\mathrm{Spec}\, K[\varepsilon]) \to \underline{A}(\mathrm{Spec}\, K)$ induced by $\varepsilon \mapsto 0$) by sending any element $g \in \underline{A}(\mathrm{Spec}\, K)$ into $g^{-1}(\underline{A}(1 + \varepsilon p))g \in \underline{A}(\mathrm{Spec}\, K[\varepsilon])$, where $1 + \varepsilon p : K[\varepsilon] \to K[\varepsilon]$ is the obvious K_o-algebra automorphism of $K[\varepsilon]$ and where we still denoted by g the image of g in $\underline{A}(\mathrm{Spec}\, K[\varepsilon])$. Next note that any morphism of functors as above $\underline{A} \to \underline{A}'$ "agrees" with $\ell_{\underline{A}}p$ and $\ell_{\underline{A}'}p$. Finally, it is not hard to check that:

1) If $\underline{A}$ is the functor of points of an algebric K_o-group then $\ell_{\underline{A}}p$ coincides with Kolchin's logarithmic derivative $\ell\nabla_{p^*}$ (to see this, use for instance the formalism in $[B_1]$ pp. 23-25).

2) If $\underline{A}'$ is the functor $S \to H^1(V_o \times S, \mathcal{O}^*)$ then $\ell_{\underline{A}'}p$ coincides with our $\ell\nabla_p$ defined in (3.1).

The discussion above plus the fact that A represents the relative Picard functor [G] closes our proof.

The following theorem establishes a description of $P(C/K_C)$ for C commutative (notations as in (1.2), (1.4), (2.1)). Intuitively it says under what conditions on $p \in P(C)$ we have that C descends to the constant field of p.

(3.4) THEOREM. Let C be an irreducible commutative algebraic K-group with abelian part A and let $p \in P(C)$. The following are equivalent:

1) p vanishes on K_C (equivalently, C is defined over $K^{K[p]}$),

2) The image of $\ell\nabla_p \in \mathrm{Hom}(H^1_{DR}(A)_m, H^1_{DR}(A)_a)$ in $\mathrm{Hom}(S_{DR}(C)_m, S(C)_a)$ vanishes

3) p vanishes on K_A and the image of the map $\ell\nabla_p \in \mathrm{Hom}(H^1(\mathcal{O}^*_A), H^1(\mathcal{O}_A))$ in $\mathrm{Hom}(S(C)_m, S(C)_a)$ vanishes.

(Of course $\ell\nabla_p$ in 2) is that defined in (1.2) while $\ell\nabla_p$ in 3) above is the one defined in (3.1)).

Proof. 2) $\Rightarrow$3). By 2) and (1.4) we get

$$\nabla_p(H^0(\Omega^1_{A/K})) \subset H^0(\Omega^1_{A/K})$$

hence by (1.5) the image $\rho(p)$ of p via the Kodaira-Spencer map vanishes. So p lifts to some derivation of $\mathcal{O}_A$ hence by (0.15) p vanishes on K_A. Now 3) follows by applying (3.2) to V = A, $K_o = K_A$.

Write in what follows $S_a = S(C)_a$.

3) $\Rightarrow$1). Since $\nabla_p(S_a) \subset S_a$, S_a is a K[p]-submodule of the split K[p]-module $H^1(\mathcal{O}_{A_o}) \otimes_{K_o} K$ (where $A = A_o \otimes_{K_o} K$ for some abelian K_o-variety, $K_o = K_A$) hence S_a itself is split: $S_a = S_o \otimes_{K_o} K$ for some K_o-subspace $S_o \subset H^1(\mathcal{O}_{A_o})$. On the other hand, upon letting $\lambda_\chi \in A^0(K)$ (where $A^0 = \mathrm{Pic}^0_{A_o/K_o}$) be the K-point corresponding to $m(\chi)$ (notation as in (2.1)), our hypothesis 3) together with Lemma (3.3) and with (I.1.11) and (1.3.6), 1) show that $\lambda_\chi \in A^0(K_o)$. This clearly implies that C is defined over K_o i.e. $K_C \subset K_o$ so $pK_C = 0$. Implications 1) $\Rightarrow$3) $\Rightarrow$2) can be proved using similar arguments.

(3.5) REMARKS. Let C be an irreducible commutative algebraic K-group. Then (with usual notaions of this chapter):

1) It may happen that neither C nor A (= abelian part of C) are defined over $K^{D(C)}$. The simplest example of this kind is that of C = E(A), cf. (2.3).

2) It may happen that A is defined over $K^{D(C)}$ but C is not. An example can be obtained as follows. We let K be a universal (ordinary) δ-field with constant field k, we let A be an elliptic curver over K admitting k as a field of definition and we let $B = G_a \times G_m$. Then let $a : X_a(B) = K \to H^1(\mathcal{O}_A)$ be any linear isomorphism and let $m : X_m(B) = \mathbf{Z} \to \mathrm{Pic}^0(A)$ be such that m(1) is taken via Kolchin's logarithmic derivative $\ell\nabla_{p^*} : \mathrm{Pic}^0 A \to H^1(\mathcal{O}_A)$ into a non-zero element (this is possible since by Kolchin's theorem $\ell\nabla_{p^*}$ is surjective!). Then using (3.3) and (3.4) one can check that the extension C of A by B described by the maps (m,a) satisfies our requirements.

3) If B (= linear part of C) is unipotent and A is defined over $K^{D(C)}$ then C itself is defined over $K^{D(C)}$. This follows directly from (3.4).

4) If B is a torus then C is defined over $K^{D(C)}$. Indeed, by (3.4) it is sufficient to check that A is defined over $K^{D(C)}$. But by (I.1.22) B is a D(C)-subgroup of C hence by (I.1.17) A becomes an algebraic D(C)-group. We conclude by (0.15).

We close by proving a regularity criterion (in Deligne's sense) for the Lie algebra of an irreducible commutative algebraic D-group.

(3.6) In the rest of this section (only) we assume (in addition to the fact that K is algebraically closed) that k is also algebraically closed, tr.deg. $K/k = m < \infty$, $P = \mathrm{Der}_k K$, $D = K[P]$.

Let V be a finite dimensional D-module: by a model of V we will understand the following data: a smooth affine k-variety X of dimension m, an embedding (over k) of the function field k(X) of X into K, a vector bundle V_X on X, an integrable connection [Del_2]

$$\nabla : V_X \to V_X \otimes \Omega^1_{X/k}$$

and a K-isomorphism $V \simeq V_X \otimes_{\mathcal{O}_X} K$ such that for any vector field v on X we have a commutative diagram

$$\begin{array}{ccc} V_X & \xrightarrow{\nabla_v} & V_X \\ \downarrow & & \downarrow \\ V & \xrightarrow[\nabla_{p(v)}]{} & V \end{array}$$

where $p(v) \in \mathrm{Der}_k K$ is the unique extension of v and $\nabla_{p(v)}$ is the multiplication with p(v) in the D-module V.

Any (finite dimensional) D-module admits a model due to the following:

(3.7) REMARK. In the situation of (3.6) if $p_1, \dots, p_m$ is a commuting K-basis of P and if $x_1, \dots, x_n \in K$ then there exists a finitely generated k-subalgebra A of K preserved by $p_1, \dots, p_m$ and containing $x_1, \dots, x_n$ (for a proof see [B_3]).

(3.8) Coming back to (3.6) we say that a finite dimensional D-module V is regular if it has a model (X, V_X, ∇) with ∇ regular in Deligne's sense [Del_2] p. 90. By loc.cit. it is easy to check that if V is regular then for any other model $(X', V'_{X'}, \nabla')$, ∇' must be regular (indeed by [Del_2] regularity is a birational concept and does not depend on passing to finite coverings). Using the latter remark and the theory in [Del_2] p. 90 it follows that:

1) K viewed as a D-module is regular.

2) If $V' \to V \to V''$ is an exact sequence of D-modules with V', V" regular then V is regular.

3) If V_1 and V_2 are regular D-modules then $V_1 \oplus V_2$, $V_1 \otimes V_2$, $\mathrm{Hom}(V_1, V_2)$ are regular, in particular V_1^o is regular.

By 1) above it follows that any split D-module (of finite dimension) is regular, since it is a sum of copies of the D-module K.

The "regularity theorem" [Del_2] p. 118 implies in particular that if V is a smooth projective K-variety then the space $H^1_{DR}(V)$ is a regular D-module, with its D-module structue given by the Gauss-Manin connection (1.1). Now here is the main consequence of our Theorem (2.2).

(3.9) THEOREM. Let C be an irreducible commutative algebraic D-group ($D = K[P]$, $P = \mathrm{Der}_k K$, tr.deg. $K/k < \infty$). Then the D-module L(C) is regular if and only if the D-module $X_a(C)$ is regular.

Proof. Let $\bar{C} = \mathrm{Spec}\, \mathcal{O}(C)$. By [DG] p. 398, $C \to \bar{C}$ is a surjective morphism of algebraic groups and its kernel G has the property that $\mathcal{O}(G) = K$. Clearly $\bar{C}$ has an induced structure of algebraic D-group and $C \to \bar{C}$ is a D-map so G is an algebraic D-subgroup and we have an exact sequence of D-modules.

(3.9.1) $\qquad 0 \to L(G) \to L(C) \to L(\bar{C}) \to 0$

Now by (I.1.22) the maximal torus $\bar{T}$ of $\bar{C}$ is an algebraic D-subgroup of $\bar{C}$ and the quotient $\bar{C}/\bar{T}$ is an algebraic vector group whose space of additive characters $X_a(\bar{C}/\bar{T})$ identifies with $X_a(C)$. It is trivial to check that the nondegenerate bilinear map

$$X_a(\bar{C}/\bar{T}) \times L(\bar{C}/\bar{T}) \to K, \qquad \langle x, v\rangle = vx$$

defined in (I.1.3) is a D-module duality (i.e. that $p\langle x,v\rangle = \langle px, v\rangle + \langle x, pv\rangle$ for $x \in X_a(\bar{C}/\bar{T})$, $v \in L(\bar{C}/\bar{T})$) so we have a D-module isomorphism $X_a(C)^o \simeq L(\bar{C}/\bar{T})$ hence an exact sequence

(3.9.2) $\qquad 0 \to L(\bar{T}) \to L(\bar{C}) \to X_a(C)^o \to 0$

Finally let T be the maximal torus of the linear part of G; by (I.1.22) it is an algebraic D-subgroup of G and the quotient $\bar{G} := G/T$ is an alebraic D-group (I. 1.17) with $\mathcal{O}(\bar{G}) = K$ and with unipotent linear part. Let A denote the abelian part of G; then $\bar{G}$ is a quotient of the universal extension E(A) hence $L(\bar{G})$ is a D-module quotient of L(E(A)). By (3.8) $H^1_{DR}(A)$ is a regular D-module hence so is its dual $H^1_{DR}(A)^o$ which by (2.3) is isomorphic to L(E(A)). By (3.8) again we conclude that $L(\bar{G})$ is a regular D-module and we have an exact sequence of D-modules

(3.9.3) $\quad 0 \to L(T) \to L(G) \to L(\bar{G}) \to 0$

Now T and $\bar{T}$ being tori are split algebraic D-groups (cf. (I. 1.9), (I. 1.26)) hence their Lie algebras L(T), $L(\bar{T})$ are split D-modules, hence are regular. Then the exact sequences (3.9.1), (3.9.2), (3.9.3) show that L(C) is regular if and only if $X_a(C)$ is so.

(3.10) REMARK. If we consider the theory which we developed in the present chapter for K a perfect field of positive characteristic then all the interesting phenomena we encountered dissappear because $Der_k K = 0$. But they reappear and in fact multiply in case K is non-perfect. We shall come back to this situation in a subsequent work.

CHAPTER 4. GENERAL ALGEBRAIC D-GROUPS

In this chapter we make a synthesis of the affine and commutative cases which were studied separately in Chapters 2 and 3 respectively. We get some results for general (non-affine, non-commutative) algebraic groups concerning local finitness (cf. section 1), automorphism functor (cf. section 2), representative ideals (cf. section 3).
Throughout this chapter we assume without any explicit mention (as we did in the previous chapter) that K is algebraically closed of characteristic zero.

1. Local finiteness criterion

(1.1) Let G be an irreducible algebraic K-group. Denote by L its linear part (= maximum linear connected subgroup of G), A its abelian part (A := G/L), Z the center of G. Recall [Ro] that G = LZ so any normal subgroup of L is normal in G. Let U be the unipotent radical of L (which will also be called "unipotent radical of G"), H a maximal reductive subgroup of L, T the radical of H and S = [H,H]. Note that G/US does not depend on the choice of H: it is the maximum semiabelian quotient of G and consequently the field K(G/US) will be called the maximum semiabelian subfield of K(G). We denote by

g, ℓ, u, h, t, s

the Lie algebras corresponding to the groups defined above. Moreover we put $\bar{G} := G/[G,G]$, $\bar{g} := g/[g,g]$, $\bar{u} := u/u \cap [g,g] \subset \bar{g}$; note that $\bar{u}$ is the Lie algebra of the unipotent radical of $\bar{G}$. Here is our main result (notations as above):

(1.2) THEOREM. Let G be an irreducible algebraic K-group.

a) The following are equivalent for $p \in P(G)$:

1) $p \in P(G, \text{fin})$

2) p preserves $\bar{u}$

3) p preserves u

4) p preserves U

5) p preserves the subfield K(G/U) of K(G)

6) p preserves the subfield K(G/US) of K(G).

b) P(G,fin) is a Lie K/k-subalgebra of P(G) and G is locally finite as a D_{fin}-variety where D_{fin} = K[P(G,fin)].

c) $K^{P(G,fin)} = K_G$

d) The natural map $P(G)/P(G,fin) \to P(\bar{G})/P(\bar{G},fin)$ is injective.

e) If $i : G \to G'$ is an isogeny and $i^* : P(G') \to P(G)$ is the lifting map then $i^*(P(G',fin)) = P(G,fin) \cap i^*(P(G'))$.

Proof. Assertions d) and e) follow from a). Assertion c) follows from b) and from (I.3.7) and (I.1.26). That P(G,fin) is a Lie K/k-subalgebra of P(G) follows from a).

Let's check 1)⇒4) in a). Indeed, by (I.3.11) upon replacing K by a K[p]-extension of K we may suppose G is a split algebraic K[p]-group. Then one easily checks that p preserves U.

Now 4)⇒3) in a) follows from (I.1.16) while 3)⇒2) is clear. Moreover 4)⇒5) follows from (I.1.17) while 5)⇒4) follows from (I.1.17) and (0.14), 5). To prove 2)⇒1) and the remaining assertion in b) it is sufficient to prove that for any Lie K/k-subalgebra P of P(G) such that $P\bar{u} \subset \bar{u}$ (or equivalently $Pu \subset u + [g,g]$) we have that G is locally finite as a D-variety, D = K[P].

Claim 1. $u + [g,g] = u + s$. Indeed $s = [s,s]$ being semisimple hence $s \subset [g,g]$ hence "⊃" follows. Conversely since G/US is semiabelian, hence commutative, $g/(u + s)$ is commutative so $[g,g] \subset u + s$ and "⊂" follows.

Claim 2. ℓ is a D-subalgebra of g. Indeed, by Claim 1 and by $Pu \subset u + [g,g]$ we get that u + s is a D-subalgebra of g. By (I.1.16) US is an algebraic D-subgroup of G. By (I.1.17) the semiabelian variety G/US becomes an algebraic D-group. By (I.1.22) the torus L/US is an algebraic D-subgroup of G/US hence $L = \Pi^{-1}(L/US)$ is an algebraic D-subgroup of G (where $\Pi : G \to G/US$ is the projection map) and we are done by (I.1.16).

Claim 3. u is a D-subalgebra of ℓ. Indeed, by Claim 1 we have $Pu \subset u + s$. On the other hand by (II.3.17) the radical of ℓ (which equals u + t) is a D-ideal in ℓ. Hence $Pu \subset (u + s) \cap$ $\cap$ (u + t) = u and our Claim is proved.

Now by (II.1.18) and (I.1.16) $\mathcal{O}(L)$ is a locally finite D-module while by (I.1.17), (I.1.9) and (0.15) A = G/L is a split algebraic D-group.

Claim 4. In order to prove that G is a locally finite D-variety it is sufficient to prove it after replacing K by some D-field extension of it K_1/K. Indeed since A is locally finite we can cover it by D-invariant open subsets A_i. Then G_i, the preimages of A_i in G, will be D-invariant affine open subsets hence by (I.3.7) $(G_i \otimes_K K_1)$ will be locally finite $K_1 \otimes_K D$-modules hence $\mathcal{O}(G_i)$ will be locally finite D-modules and the claim follows.

Now we put a "D-structure" on "Chevalley's construction" and on our construction from $[B_1]$ p. 97 of an equivariant completion of G (cf. also $[B_5]$ for group actions on the same construction). Let us quickly recall these constructions.

(1.3) One starts by choosing a finite dimensional K-subspace E of $\mathcal{O}(L)$ containing a set of generators of the maximal ideal m of $\mathcal{O}_{G,e}$ and such that E is invariant under left translations with elements in L(K). Then put $d = \dim(E \cap m)$,

$$Q = \text{projective space associated to } \overset{d}{\Lambda} E$$

$$q_0 = \text{point in } Q \text{ corresponding to } \overset{d}{\Lambda}(E \cap m)$$

and consider the induced action

$$\sigma : L \times Q \to Q, \quad \sigma(b,q) = bq$$

(the isotropy of q_0 is trivial!). This construction will be refered to as Chevalley's construction. Next we defined in $[B_1]$ p. 97 two commuting actions

$$\tau : L \times (G \times Q) \to G \times Q, \quad \tau(b,(g,q)) = (gb^{-1}, bq)$$

$$\theta : G \times (G \times Q) \to G \times Q, \quad \theta(x,(g,q)) = (xg,q)$$

and proved there is a cartesian diagram

$$\begin{array}{ccc} G \times Q & \xrightarrow{\ u\ } & Z \\ {\scriptstyle \Pi_1}\downarrow & & \downarrow{\scriptstyle w} \\ G & \xrightarrow[\ v\ ]{} & A \end{array}$$

with Z quasi-projective, Π_1 the projection onto the first factor, v the natural projection onto the abelian part and u is a principal fibre bundle with group L for the action τ. Moreover θ descends to an action $\bar{\theta} : G \times Z \to Z$ and the isotropy of $x_0 := u(e,q_0) \in Z$ with respect to $\bar{\theta}$ is still trivial. Finally, let $\overline{Gx_0} \subset Z$ be the closure of the orbit Gx_0 of x_0 under the action $\bar{\theta}$. We call $\Phi : G \to Z$ the map $\Phi(g) = \bar{\theta}(g,x_0)$; then $w \circ \Phi = v$.

(1.4) Let's introduce D-variety structures on the objects and maps occuring in (1.3). First replacing K by some D-field extension (which is allowed by Claim 4) we may assume L is split, $L \simeq L_0 \otimes K$ and taking $E = E_0 \otimes K$ with $E_0 \subset \mathcal{O}(L_0)$ big enough and L_0-invariant we may assume in (1.3) that E is a D-submodule of $\mathcal{O}(L)$. Then Q will be a D-variety, q_0 will be a D-point of Q, σ, τ and θ will be D-maps. Since Z is a geometric quotient of $G \times Q$ by τ the arguments in the proof of (I.1.17) show that Z has a natural structure of D-variety and u is a D-map. Since $e \in G$ is a D-point, (e,q_0) is a D-point of $G \times Q$ hence x_0 will be a D-point of Z. Clearly $\bar{\theta}$ is a D-map, hence so is Φ hence $\overline{Gx_0}$ will be a D-subvariety of Z. We claim that Gx_0 is a D-invariant open subset of $\overline{Gx_o}$. To check this note that $u^{-1}(Gx_o) = G \times Lq_o$ and $u^{-1}(\overline{Gx_o}) = G \times \overline{Lq_o}$ where $\overline{Lq_o}$ = closure of the τ-orbit Lq_o of q_o in $G \times Q$. Hence

$$u^{-1}(\overline{Gx_o} \setminus Gx_o) = Gx(\overline{Lq_o} \setminus Lq_o)$$

Since L is split so will be E and Q hence $\overline{Lq_o} \setminus Lq_o$ is a D-subscheme of Q. Since u is faithfully flat and affine the ideal of $\overline{Gx_o} \setminus Gx_o$ in $\mathcal{O}_Z$ is the intersection of $\mathcal{O}_Z$ with the ideal of $Gx(\overline{Lq_o} \setminus Lq_o)$ in $u_* \mathcal{O}_{GxQ}$ hence is a D-ideal and so Gx_o is D-invariant.

Now $\overline{Gx_o}$ is a projective D-variety hence by (0.15) upon replacing K by a D-field extension of it (cf. Claim 4) we may assume $\overline{Gx_o}$ is split. Since Gx_o is D-invariant it follows from (I.3.6), 4) that Gx_o (hence G which is D-isomorphic to it) is locally finite so the proof of 2) ⇒1) in (1.2), a) is finished.

(1.5) To conclude the proof of (1.2) we must check that 4) ⇒ 6). Exactly as in the proof of 4) ⇒5) it is sufficient to check that for $p \in P(G)$ we have $pu \subset u$ if and only if $p(u + s) \subset u + s$. But this can be checked via Claim 1 and the arguments used to prove Claim 3 in the proof above.

(1.6) REMARKS. 1) If in (1.2), a) we assume $p \in P(G/K)$ then condition 6) can be replaced by:

"6') p vanishes on K(G/US)"

This follows from (I.1.9).

2) Note that we implicitely reproved in (1.2) above the existence of a minimum algebraically closed field of definition for G between k and K (this was shown in [B_5]).

(1.7) COROLLARY. Let $G \to G'$ be an isogeny. Then $K_G = K_{G'}$.

Proof. Let $N = Ker(G \to G')$. To check that $K_{G'} \subset K_G$ it is sufficient to check that N and the inclusion $N \subset G$ are defined over K_G. Now N is central hence N lies in the kernel of the multiplication map $\lambda_n : Z(G) \to Z(G)$, $\lambda_n x = nx$ for some $n \geq 1$. Since Z(G) is clearly defined over K_G so is Ker λ_n (which is finite!) hence so are N and the inclusion $N \subset G$.

Conversely, to check that $K_G \subset K_{G'}$ it is sufficient by (1.2) to check that for any $p' \in P(G',fin)$ its lifting $p \in P(G)$ belongs to P(G,fin); but this follows from (1.2), a), 1)⇒3).

(1.8) COROLLARY. Let P be a Lie K/k-algebras and D = K[P]. Moreover let G be an irreducible algebraic D-group. Then

1) G is locally finite if and only if G/[G,G] is locally finite, if and only if the unipotent radical of G is an algebraic D-subgroup of G, if and only if the maximum semiabelian subfield of K<G> is a D-subfield of K<G>.

2) If $G \to G'$ is an isogeny of algebraic D-groups then G is locally finite if and only if G' is so.

2. Representing the automorphism functor

Our aim in this section is to prove the following result (cf. (I.1.23) for notations):

(2.1) THEOREM. Let G be an irreducible algebraic K-group and let $\underline{\mathrm{Aut}}\,G$: {K-schemes} → {groups} be its automorphism functor. Then:

1) The restriction of $\underline{\mathrm{Aut}}\,G$ to {reduced K-schemes} is representable by a locally algebraic K-group (call it Aut G).

2) The identity component $\mathrm{Aut}^{o}G$ is affine.

3) The kernel of the natural map Aut G → Aut L x Aut A (L = linear part of G, A = G/L) is finite.

4) The natural map λ_G : L(Aut G) → L($\underline{\mathrm{Aut}}$ G) = P(G/K) is injective and its image equals P(G/K,fin).

(2.2) REMARK. According to the general assumption of this chapter, K in the above statement is algebraically closed of characteristic zero. But if we assume K of characteristic zero not necessarily algebraically closed and also that G is not necessarily irreducible then our proof still leads to conclusions 1) and 2) in (2.1), this providing a positive answer to a question of Borel and Serre [BS] p. 152. Note that in [BS] assertion 1) was proved for linear G but much more was proved in this case namely that Aut G is of type ALA, i.e. an extension of an arithmetic group by a linear algebraic group. It was also observed there that there exist examples of non-linear (non-connected!) G's for which ($\underline{\mathrm{Aut}}$ G) (K) is "discrete" but not an extension of an arithmetic group by a finite group. In any case assertion 3) in our Theorem shows that if Aut L and Aut A are algebraic so is Aut G.

(2.3) Conclusions 1) and 2) in (2.1) are quite easy to obtain by inspecting our construction (1.3) and will be derived now, while the rest of the conclusions require further analysis. As we shall see, in way of proving assertions 3) and 4) we will get a quite precise description of all objects involved in case G is commutative.

Start with any locally algebraic reduced K-scheme Y acting on G in the sense of [BS] (i.e. with a "family" of automorphisms of G with "parameter space" Y). Then Y will also act on L and since L is linear, by [BS] $\underline{\mathrm{Aut}}$ L restricted to {reduced K-schemes} is representable by some locally algebraic K-group Aut L so there is an induced map γ : Y→ Aut L. Assume now $\gamma(Y) \subset \mathrm{Aut}^{o}L$. Under this hypothesis we can put an Y-action on the objects in (1.3). Indeed, in notations from (1.3) the space E can be chosen to be in addition $\mathrm{Aut}^{o}L$-invariant (because the natural semidirect product of L by $\mathrm{Aut}^{o}L$ acts rationally on $\mathcal{O}(L)$) so Y acts naturally on Q. One sees that our varieties $L, G, Q, Z, \overline{G} := \overline{Gx_o}$, $\overline{D} := \overline{G} \setminus \Phi(G)$ are provided then with Y-action, q_o and x_o are fixed points under these actions and all our maps $\sigma,\tau,\theta,\Pi_1,v,w,\overline{\theta},\Phi$ are Y-equivariant.

(2.4) Let us prove assertion 1) in the theorem. Since we want to apply the representability criterion in [BS] p. 140 we first construct a certain connected algebraic group H^o as follows. Let $\Gamma \subset G \times G \times G$ be the graph of the multiplication and $\overline{\Gamma}$ be the closure of Γ

in $\bar{G} \times \bar{G} \times \bar{G}$. By [G] the functor

$$S \to \{\alpha \in \mathrm{Aut}_S(\bar{G} \times S);\ \alpha(\bar{D} \times S) = \bar{D} \times S,\ \ (\alpha \times \alpha \times \alpha)(\bar{\Gamma} \times S) = \bar{\Gamma} \times S\}$$

is immediately seen to be representable on the category of locally algebraic K-schemes by a locally algebraic K-group H; we let H^o be as usual the identity component of H. There is a natural action $\eta : H^o \times G \to G$ which is faithful and hence effective in the sense of [BS] p. 139.

Let now Y be any connected reduced K-scheme of finite type acting on G and $y_o \in Y(K)$ be such that the corresponding automorphism of G is given by some $\alpha_o \in H^o$; in order for $\underline{\mathrm{Aut}}\, G$ to be representable on {reduced K-schemes} by a locally algebraic K-group Aut G (with $\mathrm{Aut}^o G = H^o$) it is sufficient by [BS] p. 140 to prove that for any $y \in Y(K)$ the corresponding automorphism of G is given by some point of $H^o(K)$. Now both H and Y act on G hence on L so by representability of $\underline{\mathrm{Aut}}\, L$ we get morphisms

$$\beta : H \to \mathrm{Aut}\, L$$

$$\gamma : Y \to \mathrm{Aut}\, L$$

Since $\gamma(y_o) = \beta(\alpha_o) \in \mathrm{Aut}^o L$ we get that $\gamma(Y) \subset \mathrm{Aut}^o L$ so our discussion in (2.3) applies. In particular Y acts on $\bar{G}$ letting $\bar{D}$ and $\bar{\Gamma}$ globally fixed so there is an induced morphism

$$\delta : Y \to H$$

Since $\delta(y_o) = \alpha_o \in H^o$ we get that $\delta(Y) \subset H^o$ and we are done.

To prove assertion 2) in the theorem, we must show that H^o is linear. Let $\tilde{G} \to \bar{G}$ be a H^o-equivariant resolution of $\bar{G}$; then the map $\nabla : \tilde{G} \to A$ is nothing but the Albanese map of $\tilde{G}$ and is H^o-equivariant (with respect to the trivial H^o-action on A). So

$$H^o \subset \mathrm{Ker}\,(\mathrm{Aut}^o \tilde{G} \to \mathrm{Aut}^o(\mathrm{Alb}(\tilde{G})))$$

which is linear by [Li] and we are done.

In fact assertion 2) can be proved more "elementary" (i.e. using neither [Li] nor "equivariant resolution"), cf. proof of assertion 3) below.

To prove assertions 3) and 4) in (2.1) we need a preparation.

(2.5) Let C be an irreducible commutative algebraic K-group and let's borrow notations from (III.2.4). Recall that $C = B' \times C^{\perp}$ with B' a vector group and $X_a(C^{\perp}) = 0$; moreover B" is the unipotent radical of $C^{\perp}$. We claim that

$$P(C/K) = (X_a(B') \otimes L(C^{\perp})) \oplus P(B'/K)$$

$$P(C/K,\mathrm{fin}) = (X_a(B') \otimes L(B'')) \oplus P(B'/K)$$

where we identify K(C) with $K(K(B') \otimes K(C^{\perp}))$. Indeed the formula for P(C/K) follows from (I.1.9). To check the second formula note that by (1.6) P(C/K,fin) consists of all $p \in P(C/K)$

vanishing on $K(C_m)$. Now since $C^{\perp} = C'' x_A C_m$ we have

$$L(C^{\perp}) = L(C'') x_{L(A)} L(C_m)$$

Let us choose a basis of $L(C^{\perp})$ as follows: first take a basis $(\alpha_i)_i$ of the image of $L(C^{\perp}) \to L(A)$ and lift it to a family $(\alpha_i'', \alpha_i^m) \in L(C'') x_{L(A)} L(C_m)$; next let $(\theta_j'')_j$ be a basis of $L(B'') = \mathrm{Ker}(L(C'') \to L(A))$ and $(\theta_k^m)_k$ be a basis of $L(B_m) = \mathrm{Ker}(L(C_m) \to L(A))$. Then:

$$((\alpha_i'', \alpha_i^m)_i\,,\ (\theta_j'')_j,\ (\theta_k^m)_k)$$

is a basis of $L(C^{\perp})$. Identifying $K(C)$ with $K(K(B') \otimes K(C'') \otimes_{K(A)} K(C_m))$ we may write any $p \in P(C/K)$ as

$$p = d \otimes 1 \otimes 1 + \Sigma f_i \otimes \alpha_i'' \otimes \alpha_i^m + \Sigma g_j \otimes \theta_j'' \otimes 1 + \Sigma h_k \otimes 1 \otimes \theta_k^m$$

where $d \in P(B'/K)$ and $f_i, g_j, h_k \in X_a(B') \subset K(B')$. Clearly, if all f_i's and h_k's vanish then p vanishes on $K(C_m)$. Conversely, assume p vanishes on $K(C_m)$. There exists a family $(x_j)_j$, $x_j \in K(A)$ with $\det(\alpha_i x_j) \neq 0$. We get $0 = px_j = \Sigma f_i \otimes (\alpha_i x_j)$ so all f_i's vanish. Finally there exist elements $(y_q)_q$, $y_q \in K(C_m)$ with $\det(\theta_k^m y_q) \neq 0$. We get $0 = py_q = \Sigma h_k \otimes (\theta_k^m y_q)$ hence $h_k = 0$ for all k and our claim is proved.

(2.6) LEMMA. In notations of (III.2.4) we have:

1) The natural map $\mathrm{Aut}\, C'' \to \mathrm{Aut}\, A$ is injective.

2) The natural map $\mathrm{Aut}\, C_m \to \mathrm{Aut}\, B_m \times \mathrm{Aut}\, A$ is injective.

3) The natural map $\mathrm{Aut}\, C^{\perp} \to \mathrm{Aut}\, B_m \times \mathrm{Aut}\, A$ is injective.

Proof. To prove 1) let $\sigma \in \mathrm{Aut}\, C''$ induce the identity on A. Viewing C'' as obtained by glueing $\mathrm{Spec}\, A_i[\xi_1,\dots,\xi_t]$, $(\xi_1,\dots,\xi_t)$ a basis of E'', we get that σ induces A-automorphisms of these rings. Let $\sigma_i \xi_k = P_{ik}(\xi)$. Compatibility with glueings gives

$$(*) \qquad P_{ik}(\xi + a_{ij}) = P_{jk}(\xi) + a_{ij}^k$$

where $\xi = (\xi_1,\dots,\xi_t)$, $a_{ij}^k = a_{ij}(\xi_k)$, $a_{ij} = (a_{ij}^1,\dots,a_{ij}^t)$.

So all P_{ik}'s (for fixed k) have the same degree n_k. If $n_k \geq 2$ for some k then Taylor's formula applied to (*) gives (writing f_i instead of P_{ik} and n instead of n_k):

$$(**) \qquad f_i^{(n)}(\xi) + \sum_k \frac{\partial f_i^{(n)}}{\partial \xi_k}(\xi) a_{ij}^k + f_i^{(n-1)}(\xi) + \dots =$$

$$= f_j^{(n)}(\xi) + f_j^{(n-1)}(\xi) + \dots + a_{ij}^k$$

where the dots stand for terms of degree at most n - 2. Looking at degree n terms in (**) we get $f_i^{(n)} = f_j^{(n)}$ so the coefficients of $(f_i^{(n)})_i$ glue together to give everywhere defined regular functions on A; consequently $f_i^{(n)} = f_j^{(n)} = f^{(n)} \in K[\xi_1,\dots,\xi_t]$. Looking at terms of degree n - 1 in

(**) we get

$$\sum_k \frac{\partial f^{(n)}}{\partial \xi_k}(\xi)a_{ij}^k = f_j^{(n-1)}(\xi) - f_i^{(n-1)}(\xi)$$

Upon making a base change in E" we may assume that $f^{(n)}$ contains ξ_1^n with coefficient 1. Letting a_i be the coefficient of ξ_1^{n-1} in $f_i^{(n-1)}$ we get (identifying coefficients of ξ_1^{n-1}):

$$na_{ij}^1 + \sum_{k\geq 2} \lambda_k a_{ij}^k = a_j - a_i$$

with $\lambda_k \in K$. But the left hand side equals $a_{ij}(n\xi_1 + \sum_{k\geq 2}\lambda_k\xi_k)$ so $n\xi_1 + \sum_{k\geq 2}\lambda_k\xi_k \in \mathrm{Ker}(a" : E" \to H^1(\mathcal{O}_A))$, contracting injectivity of a". We conclude that $n_k = 1$ for all k. The forms of degree 1 in $(P_{ik})_i$ glue together so $P_{ik} = \sum_r \lambda_{kr}\xi_r + b_{ik}$ for $\lambda_{kr} \in K$, $b_{ik} \in A_i$. Equation (*) above implies

$$(***) \qquad (\sum_r \lambda_{kr}a_{ij}^r) - a_{ij}^k = b_{jk} - b_{ik}$$

Injectivity of $E" \to H^1(\mathcal{O}_A)$ implies $\lambda_{kr} = 0$ for $r \neq k$ and $\lambda_{kk} = 1$. Hence in particular $b_{jk} = b_{ik}$ glue together to give a constant $b_k \in K$. Consequently $P_{ik} = \xi_k + b_k$ for all i and k. Since σ must have a fixed point we get $b_k = 0$ for all k so σ = identity.

To prove 2) take $\sigma \in C_m$ inducing the identity on both B_m and A. Then σ induces automorphisms σ_i of $A_i[\Sigma]$ for all i. Since for $\chi \in \Sigma$, $\sigma_i\chi$ is an invertible element of $A_i[\Sigma]$ we must have $\sigma_i\chi = a_i\chi_i$ for some $\chi_i \in \Sigma$ and some invertible element a_i of A_i. Compatibility with glueing gives

$$a_i m_{ij}(\chi_i)\chi_i = a_j m_{ij}(\chi)\chi_j$$

so $\chi_i = \chi_j$ for all i,j. Since restriction to B_m is the identity $\chi_i = \chi$ so $a_i = a_j$ for all i,j. Consequently the a_i's glue together giving a global section $\lambda \in K^*$ of $\mathcal{O}_A^*$. Clearly $\lambda = 1$ since it is so on B_m and we conclude that σ = identity.

Assertion 3) follows from 1) and 2).

(2.7) LEMMA. In notations of (III.2.4) we have

1) The map Aut C $\to$ Aut B x Aut A is injective.

2) $\mathrm{Aut}^o C = \mathrm{Hom}(B',B") \rtimes \mathrm{Aut}\, B'$ where the right hand side acts on $C = B' \times C^\perp$ by the formula $(\phi,\alpha)(x,y) = (\alpha(x), y + \phi(x))$, $\phi \in \mathrm{Hom}(B',B")$, $\alpha \in \mathrm{Aut}^o B'$, $x \in B'$, $y \in C^\perp$. In particular Theorem (2.1) holds for commutative G.

Proof. Since $X_a(C^\perp) = 0$ any isomorphism σ of $C = B' \times C^\perp$ must send $C^\perp$ onto $C^\perp$ hence sends B" onto B" and B_m onto B_m; in particular σ induces automorphisms of $C" = C^\perp/B_m$, $C_m = C^\perp/B"$ and $B' = C/C^\perp$ (call them $\sigma"$, σ_m, σ'). If σ induces the identity on A and B then by (2.6) $\sigma"$ and σ_m must be the identity, while σ' is clearly the identity. The isomorphism

$$C \simeq B' \times C'' \times_A C_m$$

shows that σ itself must be the identity and assertion 1) is proved.

Assertion 2) follows immediately from (2.5), (2.6) and the remarks above by viewing automorphisms of $B' \times C^{\perp}$ as 2 x 2 matrices whose entries are group homomorphisms.

As for the assertion that (2.1) holds in the commutative case, all we have to note is that from the assertion 2), from (2.5) and from (I.3.12), (II.3.10) we get:

$$L(\mathrm{Aut}^{o}C) = \mathrm{Hom}(L(B'), L(B'')) \oplus L(\mathrm{Aut}^{o}B') =$$

$$= (X_a(B') \otimes L(B'') \oplus P(B'/K) =$$

$$= P(C/K,\mathrm{fin})$$

Our lemma is proved.

As a by product of the proof we get:

(2.8) COROLLARY. For C an irreducible commutative algebraic K-group the cokernel of the map $\lambda_C : L(\mathrm{Aut}\, C) \to L(\underline{\mathrm{Aut}}\, C)$ identifies with $X_a(B') \otimes L(C_m)$ (notations as above).

(2.9) Let's prove assertions 3) and 4) in (2.1). Put $C = Z^{o}(G)$ so $G = LC$. If B is the linear part of C then B is of finite index in $L \cap C$ and $L/C \cap L \simeq A$. Now by (2.7) the map $\mathrm{Aut}\, G \to \mathrm{Aut}\, L \times \mathrm{Aut}(C/B)$ is injective and assertion 3) follows because $C/B \to A$ is an isogeny.

To prove 4) we proceed in several steps:

Claim 1. Any element of $\mathrm{Aut}^{o}L$ preserves $C \cap L$. Indeed $\mathrm{Aut}^{o}L$ acts on the finite group $Z(L)/Z^{o}(L)$ hence acts trivially on it. Now $C \cap L$ is an intermediate group between $Z^{o}(L)$ and $Z(L)$ which implies our claim.

Claim 2. Any element of $\mathrm{Aut}^{o}C$ preserves $C \cap L$.

Indeed $\mathrm{Aut}^{o}C$ preserves B and acts trivially on C/B hence acts trivially on $C \cap L/B$ hence preserves $C \cap L$.

Claim 3. $L(\mathrm{Aut}\, G) = L(\mathrm{Aut}\, L) \times_{L(\mathrm{Aut}(C \cap L))} L(\mathrm{Aut}\, C)$

Indeed this follows from the identification

$$\mathrm{Aut}^{o}G = \mathrm{Aut}^{o}L \times_{\mathrm{Aut}^{o}(C \cap L)} \mathrm{Aut}^{o}C$$

which in its turn follows from Claims 1 and 2.

Claim 4. $P(G/K,\mathrm{fin}) = P(L/K,\mathrm{fin}) \times_{P(C \cap L/K,\mathrm{fin})} P(C/K,\mathrm{fin})$.

Indeed, any $p \in P(G/K,\mathrm{fin})$ preserves L and C (use a splitting argument, cf. (I.3.11)) hence induces an element (p_L, p_C) of the right hand side of the latter formula. Clearly the map $p \to (p_L, p_C)$ is injective; to prove surjectivity view $G = (L \times C)/(L \cap C)$ and apply (I.1.17).

Now we conclude the proof of 4) in our Theorem. By (I.3.12) the maps

$$\lambda_L : L(\mathrm{Aut}\, L) \to P(L/K,\mathrm{fin}) \quad \text{and}$$

$$\lambda_{L \cap C} : L(\mathrm{Aut}(L \cap C)) \to P(L \cap C/K,\mathrm{fin})$$

are isomorphism (note th L ∩ C is not irreducible in general but in (I.3.12) we do not assume irreducibility!) and by (2.7) so is

$$\lambda_C : L(\mathrm{Aut}\, C) \to P(C/K,\mathrm{fin})$$

Claims 3 and 4 show that $\lambda_G : L(\mathrm{Aut}\, G) \to P(G/K,\mathrm{fin})$ must also be an isomorphism and we are done.

3. Products of abelian varieties by affine groups

As we have seen in Chapter 3 the results on P(G) do not extend from the affine to the non-affine G at all! But there is still a case when such an extension is possible namely when G is the product between its linear and its abelian part as shown by the following:

(3.1) THEOREM. Let $G = L \times A$, A an abelian K-variety and L an irreducible affine algebraic K-group. Then:

1) $K^{D(G)} = K_G$ and K_G is the algebraic closure of the compositum $K_A K_L$. In particular there is a split exact sequence of Lie K/k-algebras

$$0 \to P(G/K) \to P(G) \to \mathrm{Der}(K/K_A K_L) \to 0$$

2) $P(G/K) \simeq P(L/K) \oplus (X_a(L) \otimes L(A))$

3) $P(G/K,\mathrm{fin}) \simeq P(L/K,\mathrm{fin})$

4) P(G) contains a representative ideal. More precisely, if V is a representative ideal in P(L) (which always exists by (II.1.22)) then $V \oplus (X_a(L) \otimes L(A))$ is a representative ideal in P(G). In particular if the radical of L is unipotent then $X_a(L) \otimes L(A)$ is a representative ideal in P(G).

5) Upon identifying Aut G with Aut L x Aut A the isomorphism in 2) is equivariant (with Aut G acting on the left hand side and Aut L x Aut A acting on the right hand side naturally).

6) Upon fixing "descent" isomorphisms $A \simeq A_0 \otimes_{K_0} K$, $L \simeq L_0 \otimes_{K_0} K$, $K_0 = K_G$, letting $P_0 = \mathrm{Der}_{K_0} K$ and $D_0 = K[P_0]$ and viewing both members in assertion 2) as D_0-modules in the natural way (in the right hand side we take the D_0-module structure induced by the natural one on each factor and term) the isomorphism in 2) is an isomorphism of D_0-modules.

Proof. 1) follows from (I.1.19), (0.15), (II.1.1) and (I.1.25).

To prove 2) note that for the projection map $\Pi : G = L \times A \to L$ we have $\Pi_* \mathcal{O}_G = \mathcal{O}_L$ so K(L) is a D(G) - subfield of K(G). The restriction map $P(G/K) \to P(L/K)$ which is therefore well defined has an obvious section given by trivial lifting $p_0 \mapsto p_0 \otimes 1$. Let $p \in P(G/K)$ restrict to $0 \in P(L/K)$. Since p is a vector field on G tangent to the fibres of Π we have

$$p = \Sigma f_i \otimes \theta_i$$

where $(\theta_i)_i$ is a K-basis of $H^0(T_A)$ and $f_i \in \mathcal{O}(L)$. The equality $\mu_G p = (p \otimes 1 + 1 \otimes p)\mu_G$ is equivalent then to

$$\Sigma(\mu_L f_i - f_i \otimes 1 - 1 \otimes f_i)\mu_L \otimes \mu_A \theta_i = 0$$

Choose $(x_j)_j$, $x_j \in K(A)$ such that $\det(\theta_i x_j) \neq 0$. Applying the above equality to $1 \otimes x_j \in K(K(L) \otimes K(A))$ we get

$$\mu_L f_i - f_i \otimes 1 - 1 \otimes f_i = 0 \quad \text{for all } i$$

hence $f_i \in X_a(L)$. Conversely, if $f_i \in X_a(L)$ one immediately gets $\Sigma f_i \otimes \theta_i \in P(G/K)$; this proves 2).

3) Clearly P(L/K,fin) injects via trivial lifting $p_0 \mapsto p_0 \otimes 1$ into P(G/K,fin). Conversely, assume

$$p = d \otimes 1 + \Sigma f_i \otimes \theta_i \in P(G/K,fin)$$

with $d \in P(L/K)$, $(\theta_i)_i$ a basis of $H^0(T_A)$ and $f_i \in X_a(L)$. By (1.2) and (1.6) (and with notations from loc. cit.) p vanishes on $K((L \times A)/US) = K(K(L/US) \otimes K(A))$; this shows that d vanishes on K(L/US) hence $d \in P(L/K,fin)$. On the other hand let $x_j \in K(A)$ be such that $\det(\theta_i x_j) \neq 0$; then $0 = p(1 \otimes x_j) = \Sigma f_i \otimes \theta_i x_j$ from which we get $f_i = 0$ for all i and 3) is proved.

4) Let V be a representative ideal in P(L). Then putting

$$P_0 = Der(K/K_G)$$

$$P_1 = P(L/K,fin)$$

$$P_2 = V$$

$$P_3 = X_a(L) \otimes L(A)$$

we have $P(G) = P_0 \oplus P_1 \oplus P_2 \oplus P_3$ and the following commutation relations hold:

$$[P_i,P_j] \subset P_j \qquad \text{for all } 0 \leq i \leq j \leq 3$$

Indeed to check $[P_0,P_3] \subset P_3$ take $p \in Der(K/K_0)$, let p_G^* be the trivial lifting of p from K to $G = G_0 \otimes_{K_0} K$ ($G_0 = L_0 \times A_0$, L_0 an affine K_0-group, A_0 an abelian K_0-variety) so $p_G^* = p_L^* \otimes 1 + 1 \otimes p_A^*$ and let $v = \Sigma f_i \otimes \theta_i \in X_a(L) \otimes L(A)$. Then

$$\nabla_{p_G^*} v = [p_G^*, v] = \sum_i f_i \otimes [p_A^*, \theta_i] + \sum_i p_L^* f_i \otimes \theta_i =$$

$$= \Sigma f_i \otimes \nabla_{p_A^*} \theta_i + \Sigma \nabla_{p_L^*} f_i \otimes \theta_i \in X_a(L) \otimes L(A)$$

This computation already proves assertion 6).

To check $[P_1,P_3] \subset P_3$, $[P_2,P_3] \subset P_3$ take $v = \Sigma f_i \otimes \theta_i$ as above and $d \in P(L/K)$ (identifying it with $d \otimes 1$). Then one checks that

$$[d \otimes 1, v] = \Sigma df_i \otimes \theta_i = \Sigma \nabla_d f_i \otimes \theta_i \in X_a(L) \otimes L(A)$$

The rest of the commutation relations are clear.

So $P_2 \oplus P_3$ is an ideal in P(G) contained in P(G/K) and it is a K-Linear complement of $P(G,fin) = P_0 \oplus P_1$ in P(G). To conclude the proof of assertion 4) and to prove assertion 5) of the theorem it is sufficient to note that for any $\sigma = \sigma_L \times \sigma_A \in \operatorname{Aut} L \times \operatorname{Aut} A = \operatorname{Aut} G$ the following formulae hold:

$$\sigma_L V \sigma_L^{-1} = V$$

and for $v = \Sigma f_i \otimes \theta_i$ as above,

$$\sigma^{-1} v \sigma = (\sigma_L \otimes \sigma_A)(\Sigma f_i \otimes \theta_i)(\sigma_L^{-1} \otimes \sigma_A^{-1}) =$$

$$= \Sigma (\sigma_L f_i) \otimes (\sigma_A \theta_i \sigma_A^{-1}) \in X_a(L) \otimes L(A)$$

(where we denote as usual by $\sigma, \sigma_A \sigma_L$ the induced field automorphisms of K(G), K(A), K(L)).

(3.2) COROLLARY. Assume G is an irreducible algebraic K-group such that $G = G_1 G_2$ with G_i normal irreducible subgroups, G_1 affine and G_2 an abelian variety. Then $K^{D(G)} = K_G$. In particular there is an algebraic D-group structure on G if and only if K^D is a field of definition for G.

Proof. We have an isogeny $G_1 \times G_2 \to G$ and apply (I.1.21), (3.1) and (1.7).

EXAMPLE

We already considered in (I.1.9) and (I.1.26) the example of a product $G = A \times G_a$. With notations as in loc. cit. we have that

$$P(G/K, fin) = K\xi \frac{\partial}{\partial \xi}$$

while

$K\xi y \frac{\partial}{\partial x}$ is a representative ideal.

CHAPTER 5. APPLICATIONS TO DIFFERENTIAL ALGEBRAIC GROUPS

1. Ritt-Kolchin theory

The theory and examples of this section are classical.

(1.1) By a Δ-field we will understand in what follows a field $\mathcal{F}$ of characteristic zero equiped with a set $\Delta = \{\delta_1,\dots,\delta_m\}$ of pairwise commuting derivations $\delta_i : \mathcal{F} \to \mathcal{F}$. For $\alpha = (\alpha_1,\dots\alpha_m) \in \mathbf{N}^m$ we will write $\delta^\alpha\lambda$ instead of $\delta_1^{\alpha_1} \dots \delta_m^{\alpha_m}\lambda$ for any $\lambda \in \mathcal{F}$. If $m = 1$ we say is ordinary and we write $\lambda', \lambda'', \lambda''', \dots, \lambda^{(k)}$ instead of $\delta\lambda, \delta^2\lambda, \delta^3\lambda, \dots, \delta^k\lambda$. We define the constant field of $\mathcal{F}$ to be the field $\mathcal{F}^\Delta = \{\lambda \in \mathcal{F};\ \delta_i\lambda = 0,\ 1 \le i \le m\}$.

EXAMPLES

1) Let $(\mathcal{F}, \delta_1,\dots,\delta_m) = (\mathbf{Q}(z_1,\dots,z_m), \partial/\partial z_1,\dots,\partial/\partial z_m)$ be the field of rational functions. Here $\mathcal{F}^\Delta = \mathbf{Q}$. More generally one may take $\mathcal{F}$ any finitely generated field extension of $\mathbf{Q}$ and $\delta_i = \partial/\partial z_i$ where $z_1,\dots,z_m$ is a transcendence basis of $\mathcal{F}/\mathbf{Q}$. Note that in this case $\mathcal{F}^\Delta/\mathbf{Q}$ is algebraic. This is of course the most "geometric" example one can think of; indeed one may think $\mathcal{F}$ as the function field of some smooth affine $\mathbf{Q}$-variety V and $\delta_1,\dots,\delta_m$ as corresponding to pairwise commuting vector fields on V generating its tangent bundle.

2) Let $(\mathcal{F}, \delta_1,\dots,\delta_m) = (\mathbf{Q}(z_1,\dots,z_m, \exp z_1,\dots,\exp z_m), \partial/\partial z_1,\dots,\partial/\partial z_m)$. Here $\mathcal{F}^\Delta = \mathbf{Q}$ too. This should also be viewed as a "geometric" example, in the sense that it "comes from" m commuting vector fields on a variety V which is a model of $\mathcal{F}/\mathbf{Q}$; we can be very specific here by taking $V = \operatorname{Spec} \mathbf{Q}[z_1,\dots,z_m,t_1,\dots,t_m]$ and letting the vector fields be

$$\frac{\partial}{\partial z_1} + t_1\frac{\partial}{\partial t_1}, \dots, \frac{\partial}{\partial z_m} + t_m\frac{\partial}{\partial t_m}$$

3) Let $\mathcal{F} = \mathbf{Q}\langle y\rangle = \mathbf{Q}(y_0, y_1, y_2, \dots)$ be the field of rational functions in infinitely many variables $y_0, y_1, y_2, \dots$ equiped with the derivation $\delta : \mathcal{F} \to \mathcal{F}$ defined by $\delta y_i = y_{i+1}$, $i \ge 0$. This example is very different in nature from Examples 1, 2 because y_0 and its derivatives are algebraically independent over $\mathcal{F}^\Delta = \mathbf{Q}$; it should be viewed as very "non-geometric". Nevertheless it will be crucial to include this kind of examples in the theory!

4) Let $(\mathcal{F}, \delta_1,\dots,\delta_m) = (\mathrm{Mer}(R), \partial/\partial z_1,\dots,\partial/\partial z_m)$ where Mer(R) is the field of meromorphic functions on the domain $R \subset \mathbf{C}^m$. Here $\mathcal{F}^\Delta = \mathbf{C}$.

5) Let $\mathcal{F} = \mathrm{Mer}(\mathbf{C}^m, 0)$ be the quotient field of the ring $\mathcal{O}^{an}_{\mathbf{C}^m,0}$ of germs of holomorphic functions at $(\mathbf{C}^m, 0)$ and $\delta_i = \partial/\partial z_i$, $1 \le i \le m$.

(1.2) By a morphism of Δ - fields we will understand a field homomorphism $\sigma : \mathcal{F} \to \mathcal{G}$ such that

$$\sigma(\delta_i\lambda) = \delta_i(\sigma\lambda)$$

for all $\lambda \in \mathcal{F}$ and $1 \leq i \leq m$. If σ is an inclusion we say that $\mathcal{G}$ is a Δ-extension of $\mathcal{F}$ and that $\mathcal{F}$ is a Δ-subfield of $\mathcal{G}$. We say that $\mathcal{G}$ is Δ - finitely generated over $\mathcal{F}$ if there exist $\lambda_1,\ldots,\lambda_n \in \mathcal{G}$ such that $\mathcal{G}$ is generated as a field extension of $\sigma\mathcal{F}$ by the family $(\delta^{\alpha}\lambda_i)_{1\leq i\leq n, \alpha\in\mathbf{N}^m}$.

EXAMPLES

We have the following natural morphisms of Δ-fields

$$Q \xrightarrow{\sigma_1} Q(z_1,\ldots,z_m) \xrightarrow{\sigma_2} Q(z_1,\ldots,z_m,\exp z_1,\ldots,\exp z_m) \to \mathrm{Mer}(\mathbf{C}^m) \to \mathrm{Mer}(\mathbf{C}^m,0)$$

$$Q \xrightarrow{\sigma_3} Q\langle y\rangle \to \mathrm{Mer}(\mathbf{C},0), \quad y_i \mapsto d^i\phi/dz^i$$

where $\phi \in \mathrm{Mer}(\mathbf{C},0)$ has the property that the family consisting of ϕ and its derivatives of arbitrary order are algebraically independent over Q (such functions are called transcendentally transcendental and their existence is easily established). Of the above Δ-morphisms $\sigma_1\sigma_2\sigma_3$ are the ones which provide Δ-finitely generated Δ-extensions.

Other remarkable examples of morphisms of Δ-fields are provided by restrictions

$$\mathrm{Mer}(R) \to \mathrm{Mer}(R_1)$$

where R_1 is a subdomain of R.

(1.3) Let $\mathcal{G}$ be a Δ-extension of $\mathcal{F}$; one says that $\mathcal{G}$ is semiuniversal over $\mathcal{F}$ if for any morphism of Δ-fields $\sigma : \mathcal{F} \to \mathcal{F}_1$ with $\mathcal{F}_1$ Δ-finitely generated over $\mathcal{F}$ there exists a morphism $\tau : \mathcal{F}_1 \to \mathcal{G}$ making the following diagram commutative

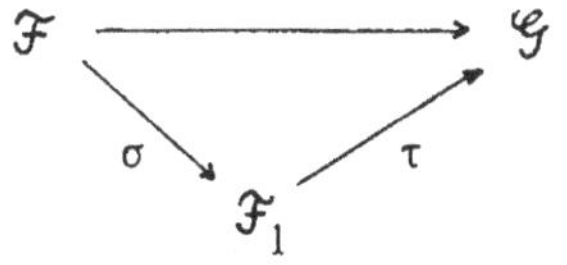

A basic result of Seidenberg [Sei] says that Mer(R) is a semiuniversal extension of **Q** if R, say, is a disk in **C**.

(1.4) Let $\mathcal{U}$ be a Δ-extension of $\mathcal{F}$; one says $\mathcal{U}$ is universal over $\mathcal{F}$ if $\mathcal{U}$ is semiuniversal over any Δ-finitely generated extension of $\mathcal{F}$ contained in $\mathcal{U}$.

Such an $\mathcal{U}$ is always algebraically closed hence so is its constant field.

A Δ-field will be called universal if it is universal over **Q**.

A basic result of Kolchin [K_1] says that any Δ-field $\mathcal{F}$ has a universal Δ-extension $\mathcal{U}$ (which is countable if $\mathcal{F}$ is so). On the other hand Seidenberg's results in [Sei] imply that any countable Δ-field can be Δ-embedded into $\mathrm{Mer}(\mathbf{C}^m,0)$. In particular $\mathrm{Mer}(\mathbf{C}^m,0)$ contains Δ-subfields $\mathcal{U}$ which are universal; admittedly no explicit description of such an $\mathcal{U}$ seems to be available. But at least we dispose of "theoretic" examples of universal Δ-fields which are realized as Δ-fields of meromorphic functions.

(1.5) By a Δ-$\mathcal{F}$-algebra over a Δ-field $\mathcal{F}$ one understands an $\mathcal{F}$-algebra $\mathcal{R}$ equiped with m commuting derivations on $\mathcal{R}$ lifting $\delta_1,\ldots,\delta_m$ (and still denoted by $\delta_1,\ldots,\delta_m$). By a Δ-ideal in $\mathcal{R}$ one understands an ideal $J \subset \mathcal{R}$ such that $\delta_i(J) \subset J$, $1 \leq i \leq m$. If $S \subset \mathcal{R}$ is a subset one denotes by (S) the smallest ideal of $\mathcal{R}$ containing S, by [S] the smallest Δ-ideal of $\mathcal{R}$ containing S and by {S} the smallest perfect Δ-ideal of $\mathcal{R}$ containing S. Note that $\{S\} = \{r \in \mathcal{R};\ \exists\ n$ such that $r^n \in [S]\}$. There is an ambiguity of notation here: if $x_1,\ldots,x_p \in \mathcal{R}$ then $\{x_1,\ldots,x_p\}$ will generally denote the smallest perfect Δ-ideal of $\mathcal{R}$ containing $x_1,\ldots,x_p$ rather than the set whose elements are $x_1,\ldots,x_p$ (unless the contrary is obvious from context).

A Δ-$\mathcal{F}$-algebra $\mathcal{R}$ is called Δ-finitely generated if there exist $x_1,\ldots,x_n \in \mathcal{R}$ such that is generated as an algebra by the family $(\delta^{\alpha} x_i)_{1 \leq i \leq n, \alpha \in \mathbf{N}^m}$.

By a Δ-$\mathcal{F}$-algebra map we understand an $\mathcal{F}$-algebra map which commutes with the derivations in the obvious way.

EXAMPLE

Let $\mathcal{R} := \mathcal{F}\{y_1,\ldots,y_n\}$ denote the $\mathcal{F}$-algebra of polynomials in the indeterminates $y_i^{(\alpha)}$ where $1 \leq i \leq n$, $\alpha \in \mathbf{N}^m$ viewed as a Δ-$\mathcal{F}$-algebra with the derivations $\delta_j : \mathcal{R} \to \mathcal{R}$ satisfying

$$\delta^{\beta} y_i^{(\alpha)} = y_i^{(\alpha+\beta)}$$

for all $1 \leq i \leq n$, $\alpha,\beta \in \mathbf{N}^m$. We write y_i instead of $y_i^{(o)}$ and if $m = 1$ (ordinary case) we write y', y'', y''' instead of $y^{(1)}, y^{(2)}, y^{(3)}$. So for instance if $m = 1$, $n = 2$ and if $\lambda \in \mathcal{F}$ we have in $\mathcal{F}\{y_1,y_2\}$ the equality:

$$\delta(\lambda y_1 y_2' - (y_1'')^2) = \lambda' y_1 y_2' + \lambda y_1' y_2' + \lambda y_1 y_2'' - 2 y_1'' y_1'''$$

One calls $\mathcal{F}\{y_1,\ldots,y_n\}$ the ring of Δ-polynomials in the Δ-indeterminates $y_1,\ldots,y_n$.

To illustrate the concept of Δ-ideal start with a Δ-polynomial $F \in \mathcal{F}\{y_1,\ldots,y_n\}$ where is, say, ordinary. Then the ideal [F] can be described of course as

$$[F] = (F, \delta F, \delta^2 F, \ldots, \delta^k F, \ldots)$$

but the ideal {F} does not have in general such an easy description by generators. We make the important remark that even if F is irreducible (as a polynomial) the ideal {F} may fail to be prime. E.g. take $n = 1$, $F \in \mathcal{F}\{y\}$, $F = (y'')^2 - y$, [Ri] p. 24; then the ideal {F} is not prime because combining $\delta^2 F$ and $\delta^3 F$ we get

$$y''(4y^{(3)}y^{(5)} - 12(y^{(4)})^2 + 8y^{(4)} - 1) \in [F]$$

while none of the factors above belongs to {F}.

Concerning Δ-ideals of Δ-polynomials we have the following basic results due to Ritt:

(1.6) THEOREM ([Ri] p. 10). For any perfect Δ-ideal J in $\mathcal{F}\{y_1,\dots,y_n\}$ there is a finite subset $S \subset J$ such that $J = \{S\}$.

(1.7) THEOREM ([Ri] p. 13). Any perfect Δ-ideal in $\mathcal{F}\{y_1,\dots,y_n\}$ is a finite intersection of prime Δ-ideals.

The problem of making (1.6) and (1.7) as "explicit" as possible was treated in detail by Ritt [Ri].

(1.8) From now on we fix throughout this Chapter a universal Δ-field $\mathcal{U}$ and we denote by $\mathcal{K} = \mathcal{U}^\Delta$ its constant field. Call the set $\mathcal{U}^n$ the n-affine space. For any subset S of $\mathcal{U}\{y_1,\dots,y_n\}$ we define the zero locus of S in $\mathcal{U}^n$ by the formula

$$Z(S) = \{\alpha \in \mathcal{U}^n;\ F(\alpha) = 0 \text{ for all } F \in S\}$$

The subsets of $\mathcal{U}^n$ of the form $Z(S)$ as above will be called Δ-closed subsets. Conversely for any subset Σ of $\mathcal{U}^n$ we define the ideal of Σ as follows:

$$I(\Sigma) = \{ f \in \mathcal{U}\{y_1,\dots,y_n\};\ F(\alpha) = 0 \text{ for all } \alpha \in \Sigma\}$$

Clearly $I(\Sigma)$ is a perfect Δ-ideal in $\mathcal{U}\{y_1,\dots,y_n\}$. Due to universality of $\mathcal{U}$ the correspondence

$$S \to Z(S)$$

$$\Sigma \to I(\Sigma)$$

establishes a bijection between Δ-closed subsets of $\mathcal{U}^n$ and perfect Δ-ideals in $\mathcal{U}\{y_1,\dots,y_n\}$. The Δ-closed subsets of $\mathcal{U}^n$ form a topology called the Δ-topology; this topology is stronger than the Zariski topology and by (1.6) it is a Noetherian topology. A Δ-closed subset Σ is irreducible iff $I(\Sigma)$ is prime.

EXAMPLES

1) $\mathcal{K}^n$ is an irreducible Δ-closed subset of $\mathcal{U}^n$; indeed $\mathcal{K}^n = Z(y'_1,\dots,y'_n)$ and $I(\mathcal{K}^n) = [y'_1,\dots,y'_n]$.

2) The Δ-closed subset $Z((y'')^2 - y)$ of $\mathcal{U}$ is reducible.

3) The Δ-closed subset $Z(y^2 - y'')$ of $\mathcal{U}$ is irreducible.

(1.9) Let Σ be an irreducible Δ-closed subset of $\mathcal{U}^n$; then Σ has the topology induced from the Δ-topology on $\mathcal{U}^n$ which we call the Δ-topology of Σ. Let Ω be a Δ-open subset of Σ (i.e. an open set for the Δ-topology); a function $f: \Omega \to \mathcal{U}$ will be called Δ-regular at $x_0 \in \Omega$ if there exist a Δ-open neighbourhood Ω_0 of x_0 in Ω and Δ-polynomials $F, G \in \mathcal{U}\{y_1,\dots,y_n\}$ with G nowhere zero on Ω_0 and

$$f(x) = \frac{F(x)}{G(x)}$$

for all $x \in \Omega_0$; f will be called Δ-regular on Ω if it is Δ-regular at any point of Ω. Any Δ-regular function $\Omega \to \mathcal{U}$ is continuous with respect to the Δ-topologies on Ω and $\mathcal{U}$. Define

$$(\Omega) := \{f : \Omega \to \mathcal{U}; \text{ f is } \Delta\text{-regular on } \Omega\}$$

It is a Δ-$\mathcal{U}$-subalgebra of the Δ-$\mathcal{U}$-algebra of all functions $\Omega \to \mathcal{U}$. In particular we dispose of the Δ-$\mathcal{U}$-algebra $\mathcal{O}(\Sigma)$ of all Δ-regular functions $\Sigma \to \mathcal{U}$. Note on the other hand that we dispose of the algebra

$$\mathcal{U}\{\Sigma\} := \mathcal{U}\{y_1, \dots, y_n\}/I(\Sigma)$$

which can be called the Δ-coordinate algebra of Σ and that we dispose of an injective Δ-$\mathcal{U}$-algebra map

$$\mathcal{U}\{\Sigma\} \to \mathcal{O}(\Sigma)$$

This map may not be surjective (and this is a crucial contrast with algebraic geometry!). Nevertheless it is easily seen to induce an isomorphism to the quotient fields. The common quotient field of $\mathcal{U}\{\Sigma\}$ and $\mathcal{O}(\Sigma)$ is denoted by $\mathcal{U}\langle\Sigma\rangle$. Define $\dim \Sigma = \text{tr.deg. } \mathcal{U}\langle\Sigma\rangle/\mathcal{U}$.

EXAMPLES

1) If $\Sigma = \mathcal{U}^n$ then $\mathcal{U}\{\Sigma\} \to \mathcal{O}(\Sigma)$ is easily seen to be an isomorphism; in other words $\mathcal{O}(\mathcal{U}^n) = \mathcal{U}\{y_1, \dots, y_n\}$. Note that $\dim \mathcal{U}^n = \infty$. More generally (but this is more subtle) we have shown in [B_8] that $\mathcal{U}\{\Sigma\} \to \mathcal{O}(\Sigma)$ is an isomorphism if Σ is Zariski closed in $\mathcal{U}^n$ and, as an algebraic variety, is smooth. We will not use this result in what follows.

2) Let $\Sigma = \mathcal{K} = Z(y') \subset \mathcal{U}$. Then if $\lambda \in \mathcal{U}$, $\lambda \notin \mathcal{K}$ we have that $1/(\eta - \lambda) \in \mathcal{O}(\Sigma)$, $1/(\eta - \lambda) \notin \mathcal{U}\{\Sigma\}$ where η is the image of y in $\mathcal{U}\{\Sigma\}$. Note that $\dim(\mathcal{K}) = 1$.

(1.10) Let F be any field. By a naïve ringed space over F we will understand a pair $(X, \mathcal{O})$ consisting of a topological space X and of a subsheaf $\mathcal{O}$ of F-algebras of the sheaf F_X of all F-valued functions on the open sets of X. By a morphism $(X, \mathcal{O}_X) \to (Y, \mathcal{O}_Y)$ of naïve ringed spaces over F we will understand a continuous map $f : X \to Y$ with the property that for any open set $V \subset Y$ and any $\phi \in \mathcal{O}_Y(V)$ we have $\phi \circ f \in \mathcal{O}_X(f^{-1}(V))$.

We obtained a category

$$\{\text{naïve ringed spaces over F}\}$$

(1.11) By an affine Δ-manifold we will understand an object in the category of naïve ringed spaces over $\mathcal{U}$ isomorphic to $(\Sigma, \mathcal{O}_\Sigma)$ where Σ is an irreducible Δ-closed subset of some $\mathcal{U}^n$ and $\mathcal{O}_\Sigma$ is the sheaf of Δ-regular functions:

$$\Omega \to \mathcal{O}(\Omega)$$

defined in (1.9). By a Δ-manifold we will understand a naïve ringed space $(\Sigma, \mathcal{O})$ over $\mathcal{U}$ which is irreducible and can be covered by finitely many open sets $\Sigma_1,\dots,\Sigma_p$ such that each $(\Sigma_i, \mathcal{O}_{|\Sigma_i})$ is an affine Δ-manifold. For any Δ-manifold Σ we define $\mathcal{U}\langle\Sigma\rangle$ as the direct limit of all $\mathcal{O}_\Sigma(\Omega)$ for $\Omega \subset \Sigma$ non-empty Δ-open sets; then $\mathcal{U}\langle\Sigma\rangle$ is a Δ-extension of $\mathcal{U}$ (and in case Σ is affine it coincides with the one defined at (1.9)). We define the dimension of Σ by

$$\dim \Sigma := \text{tr.deg. } \mathcal{U}\langle\Sigma\rangle/\mathcal{U}.$$

By a morphism of Δ-manifolds we simply understand a morphism of naïve ringed spaces over $\mathcal{U}$. We obtained a category

$$\{\Delta\text{-manifolds}\}$$

We leave to the reader the verification that this category has direct products. As expected if $\Sigma \subset \mathcal{U}^n$, $\Gamma \subset \mathcal{U}^m$ are Δ-closed irreducible subsets then $\Sigma \times \Gamma \subset \mathcal{U}^{n+m}$ is Δ-closed and irreducible and so it has a structure of affine Δ-manifold; this construction is "invariant" and "globalizes" via "glueing" to get the existence of products in the expected way. A group object in the category of Δ-manifolds will be called a Δ-group; so a Δ-group is a Δ-manifold Γ equiped with morphisms

$$\mu : \Gamma \times \Gamma \to \Gamma$$

$$S : \Gamma \to \Gamma$$

and with an element $e \in \Gamma$ satisfying the obvious group axioms. With the obvious notion of morphisms of Δ-groups we obtain a category

$$\{\Delta\text{-groups}\}$$

A Δ-manifold of finite dimension will be called a Δ_0-manifold.

A Δ-group of finite dimension will be called a Δ_0-group. Any Δ-group in our sense "provides" an "irreducible Δ-group" in Kolchin's sense [K$_2$]; it is not clear whether the converse holds. One can prove that our Δ_0-groups "are" precisely Kolchin's irreducible Δ-groups of type zero [K$_2$].

Before giving a few examples of Δ-manifolds and Δ-groups it is convenient to discuss more generalities. Let F be an algebraically closed field. In the rest of this section (only) F-varieties will always be viewed as naïve ringed spaces (rather than schemes) over F; so they will be identified with their sets of F-points and their structure sheaf will be the sheaf of (F-valued) regular functions. F-varieties will always be assumed irreducible.

(1.12) Let X be an affine $\mathcal{U}$-variety and take an embedding $i : X \to \mathbf{A}^n = \mathcal{U}^n$; then X being Zariski closed in $\mathcal{U}^n$ is in particular Δ-closed and by a basic result of Kochin [K$_1$] is irreducible in the Δ-topology so it has by (1.9), (1.11) a structure of affine Δ-manifold which we call $\hat{X}$. We leave to the reader the task of checking that this structure does not depend on the embedding i. Note that $\dim \hat{X} = \infty$ if $\dim X > 0$.

Now if X is any $\mathcal{U}$-variety, cover X by affine Zariski open sets X_i and note that the Δ-manifolds $\hat{X}_i$ glue together to give a Δ-manifold structure on X which we call $\hat{X}$; $\hat{X}$ will be called the Δ-manifold produced from X. As sets we have $\hat{X} = X$ but the topology on $\hat{X}$ (called the Δ-topology) is stronger than the (Zariski) topology of X. The Δ-manifold $\hat{X}$ does not depend on the affine covering of X and is "naturally" associated to X.

Actually one may check that $X \mapsto \hat{X}$ gives a functor

$$\{\mathcal{U}\text{-varieties}\} \to \{\Delta\text{-manifolds}\}$$

It worths noting that if X, Y are $\mathcal{U}$-varieties then the map

$$\mathrm{Hom}_{\mathcal{U}\text{-var}}(X,Y) \to \mathrm{Hom}_{\Delta\text{-man}}(\hat{X},\hat{Y})$$

is injective but not surjective in general. Indeed if $X = Y = \mathbf{A}^1$ then $\mathrm{Hom}_{\Delta\text{-man}}(\hat{\mathbf{A}}^1, \hat{\mathbf{A}}^1)$ contains the map $f(y) = (y')^2 - y^3$, say, which does not belong to $\mathrm{Hom}_{\mathcal{U}\text{-var}}(\mathbf{A}^1,\mathbf{A}^1)$. The functor above commutes with products so induces a functor $G \mapsto \hat{G}$

$$\{\text{irreducible algebraic } \mathcal{U}\text{-groups}\} \to \{\Delta\text{-groups}\}$$

So we already dispose of a series of examples of Δ-manifolds and Δ-groups :

EXAMPLES

For any $\mathcal{U}$-variety X we dispose of a Δ-manifolds $\hat{X}$; in particular $\hat{\mathbf{A}}^n$, $\hat{\mathbf{P}}^n$ are Δ-manifolds. Any irreducible Δ-closed subset Σ of a Δ-manifold $\hat{X}$ has a structure of Δ-manifold (the way of seeing this is rather trivial and we leave it to the reader). In particular the Δ-closed sets below have natural structures of Δ-manifolds:

1) $Z(y'' + ay' + by) \subset \mathcal{U} = \hat{\mathbf{A}}^1$, $a,b \in \mathcal{U}$, y a Δ-indeterminate.

2) $Z(xy - 1, yy'' - (y')^2 + ayy') \subset \mathcal{U}^2 = \hat{\mathbf{A}}^2$, $a \in \mathcal{U}$, x,y two Δ-indeterminates.

3a) $Z(y^2 - x(x-1)(x-c), x''y - x'y' + ax'y) \subset \mathcal{U}^2 = \hat{\mathbf{A}}^2$, $c \in \mathcal{K}$, $a \in \mathcal{U}$, x,y as above.

3b) $Z(y^2 - x(x-1)(x-t), -y^3 - 2(2t-1)(x-t)^2x'y + 2t(t-1)(x-t)^2(x''y - 2x'y')) \subset \mathcal{U}^2 = \hat{\mathbf{A}}^2$, $t \in \mathcal{U}$, $t' = 1$, x,y as above.

4) $Z(y_{12},y_{21},y_{31},y_{32},y_{22} - 1,y_{33} - 1,y'_{23},y'_{11} - y_{23}y_{11}) \subset \widehat{GL}_3$, y_{ij} Δ-indeterminates $1 \leq i,j \leq 3$.

The irreducibility of these sets is easily established directly in all cases except 3a) and 3b); the latter cases are more subtle but nevertheless they can be treated using Ritt's theory and we leave them to the reader, too. Examples 1), 2), 3) have finite dimension while 4) has infinite dimension.

Now for any irreducible algebraic $\mathcal{U}$-group G we dispose of a Δ-group $\hat{G}$; in particular $\hat{G}_a$, $\hat{G}_m$, $\widehat{GL}_N$, $\hat{A}$ are Δ-groups (of infinite dimension), where A is any non-zero abelian $\mathcal{U}$-variety. It is easy to check that any irreducible Δ-closed subgroup of $\hat{G}$ has a natural structure of Δ-group. On the other hand if Σ is one of the Δ-closed subsets defined in 1)-4) above then the following hold. If Σ is given by 1) then Σ a Δ-closed subgroup of $\hat{\mathbf{A}}^1 = \hat{G}_a$. If Σ is given by 2) then Σ is a Δ-closed subgroup of $Z(xy - 1) = \hat{G}_m \subset \hat{\mathbf{A}}^2$. If Σ is given by 4) then Σ is a Δ-closed subgroup of $\hat{GL}_3$. Finally if Σ is given by 3a) respectively 3b) then the Δ-closure Γ of Σ in $\hat{\mathbf{P}}^2$ is a subgroup of $\hat{A} \subset \hat{\mathbf{P}}^2$ where $A \subset \mathbf{P}^2$ is the elliptic curve obtained by taking the Zariski closure of $Z(y^2 - x(x-1)(x-c))$ respectively $Z(y^2 - x(x-1)(x-t))$ in $\mathbf{P}^2$. All these statements can be checked by direct computation! We conclude that in examples 1), 2) Σ is a Δ_o-group, in example 4) Σ is a Δ-group while in examples 3a), 3b) Γ is a Δ_o-group. As remarked in the Introduction examples 1), 2), 3a), 4) are due to Cassidy and Kochin while example 3b) is implicit in Manin's paper [Ma]. Here is another remarkable class of examples of Δ-groups of infinite dimension due to Cassidy an related to work of Ritt:

5) $\Gamma = \hat{\mathbf{A}}^2$, $\mu : \Gamma \times \Gamma \to \Gamma$, $\mu((u_1,u_2),(v_1,v_2)) = (u_1 + v_1, u_2 + v_2 + \sum_{i<j} a_{ij} u_1^{(i)} v_1^{(j)})$, $a_{ij} \in \mathcal{U}$.

Many other examples may be found in Casidy's papers $[C_i]$.

(1.13) In (1.12) we constructed Δ-manifolds starting with $\mathcal{U}$-varieties. Now we shall construct Δ-manifolds starting with $\mathcal{K}$-varieties.

Let X_o be an affine $\mathcal{K}$-variety, choose an embedding $i_o : X_o \to \mathcal{K}^n$ and compose it with the natural inclusion $\mathcal{K}^n \to \mathcal{U}^n$; then X_o appears as a Δ-closed subset of $\mathcal{U}^n$ (indeed if X_o is the zero set in $\mathcal{K}^n$ of $F_1,\dots,F_p \in \mathcal{K}[y_1,\dots,y_n]$ then X_o is the zero set in $\mathcal{U}^n$ of $F_1,\dots,F_p$, $y'_1,\dots,y'_n \in \mathcal{U}\{y_1,\dots,y_n\}$). So X_o has a natural structure of affine Δ-manifold, cf. (1.9), (1.11) which we call $\check{X}_o$ and which is easily seen to be independent of the embedding i_o.

Now if X_o is any $\mathcal{K}$-variety we may cover X_o with affine Zariski open sets X_{oi} and note that the Δ-manifolds $\check{X}_{oi}$ glue together to give a Δ_o-manifold $\check{X}_o$ with $\dim \check{X}_o = \dim X_o$. As topological spaces $\check{X}_o = X_o$ but the structure sheaves are drastically different (e.g. if $X_o = \mathcal{K}$ we already saw that $\mathcal{O}(\check{X}_o)$ contains the function $f : \mathcal{K} \to \mathcal{U}$, $f(\eta) = 1/(\eta - \lambda)$ where $\lambda \in \mathcal{U}$, $\lambda \notin \mathcal{K}$; but of course f is $\mathcal{U}$-valued, not $\mathcal{K}$-valued so $f \notin \mathcal{O}(X_o)$.

We obtained a functor $X_o \mapsto \check{X}_o$

$$\{\mathcal{K}\text{-varieties}\} \to \{\Delta_o\text{-manifolds}\}$$

inducing a functor $G_o \mapsto \check{G}_o$

$$\{\text{irreducible algebraic } \mathcal{K}\text{-groups}\} \to \{\Delta_o\text{-groups}\}$$

Any Δ_o-group isomorphic to $\check{G}_o$ for some G_o as above will be called split. Finding criteria for a Δ_o-group to be split will be one of our main concerns.

2. Δ_o-groups versus algebraic D-groups

(2.1) Recall that we have fixed in the preceeding section a universal Δ-field $\mathcal{U}$ with constant field $\mathcal{K}$. Then let P be the free integrable Lie $\mathcal{U}/\mathcal{K}$-algebra built on $\delta_1,\dots,\delta_m$ (0.5); recall that $P = \mathcal{U}\delta_1 \oplus \dots \oplus \mathcal{U}\delta_m$ and

$$[\lambda\delta_i,\mu\delta_j] = \lambda(\delta_i\mu)\delta_j - \mu(\delta_j\lambda)\delta_i$$

for all i,j and $\lambda, \mu \in \mathcal{U}$. Moreover let $D = \mathcal{U}[P]$; recall that D has an $\mathcal{U}$-basis consisting of monomials δ^α, $\alpha \in \mathbf{N}^m$, i.e. $D = \sum_{\alpha\in\mathbf{N}^m} \mathcal{U}\delta^\alpha$, with the "obvious" $\mathcal{K}$-algebra structure. With these notations remark that the concept of Δ-extension of $\mathcal{U}$ in (1.2) is the same with the concept of D-field in (0.5). The concept of Δ-$\mathcal{U}$-algebra in (1.5) is the same with that of D-algebra in (0.5). The concept of Δ-ideal in a Δ-$\mathcal{U}$-algebra in (1.5) is the same with the concept of D-ideal in a D-algebra, cf. (0.7). A Δ-$\mathcal{U}$-algebra is Δ-finitely generated in the sense of (1.5) iff the corresponding D-algebra is D-finitely generated in the sense of (0.5). By (1.6) and (1.7) $\mathcal{U}$ is D-algebraically closed in the sense of (0.5).

By the above discussion, any integral D-scheme of finite type can be covered by D-schemes of the form Spec $\mathcal{U}\{y_1,\dots,y_n\}/J$ where J is some prime Δ-ideal.

(2.2) In order to relate the Ritt-Kolchin frame (cf. Section 1 of this Chapter) to our frame in Chapter 1-4 it will be convenient to introduce a basic functor

$$\{\text{integral D-schemes of D-finite type}\} \to \{\Delta\text{-manifolds}\},\quad V \mapsto V_\Delta$$

This is done as follows. Assume V is an integral D-scheme of D-finite type; then as a set we put

$$V_\Delta = \mathrm{Hom}_{D\text{-}sch}(\mathrm{Spec}\,\mathcal{U}, V)$$

Clearly V_Δ identifies with the subset of V consisting of all $p \in V$ for which the maximal ideal m_p of $\mathcal{O}_{V,p}$ is a Δ-ideal and the residue field $\mathcal{O}_{V,p}/m_p$ is a trivial extension of $\mathcal{U}$. In notations of (0.7) we actually have $V_\Delta = V_D(\mathcal{U})$. The subsets of V_Δ of the form $U_\Delta = \mathrm{Hom}_{D\text{-}sch}(\mathrm{Spec}\,\mathcal{U}, U)$, where U is any Zariski open subset of V, are the open sets of a topology on V_Δ which we call the Δ-topology. For any U as above we put

$$\mathcal{O}(U_\Delta) = \bigcap_{p\in U_\Delta} \mathcal{O}_{V,p}$$

Any element f in $\mathcal{O}(U_\Delta)$ can be identified with a function $f : U_\Delta \to \mathcal{U}$ via the formula $f(p) = (f \bmod m_p) \in \mathcal{U}$. So we obtained a subsheaf

$$U_\Delta \mapsto \mathcal{O}(U_\Delta)$$

to of the sheaf of all $\mathcal{U}$-valued functions on V_Δ. Call this subsheaf $\mathcal{O}_{V_\Delta}$; then we leave to the reader the task to check that $(V_\Delta, \mathcal{O}_{V_\Delta})$ is a Δ-manifold and that actually we got a functor from {integral D-schemes of D-finite type} to {Δ-manifolds}. The key point to be checked is actually the fact that if $V = \mathrm{Spec}\,\mathcal{U}\{y_1,\dots,y_n\}/J$ for some prime Δ-ideal J then

$$(V_\Delta, \mathcal{O}_{V_\Delta}) \simeq (\Sigma, \mathcal{O}_\Sigma) \text{ as naive ringed spaces over } \mathcal{U}$$

where $\Sigma = Z(J) \subset \mathcal{U}^n$ is the zero set of J and $\mathcal{O}_\Sigma$ is the sheaf of Δ-regular functions on Σ (this shows in particular that any affine Δ-manifold has the form $(\mathrm{Spec}\, R)_\Delta$ where R is an integral Δ-finitely generated Δ-$\mathcal{U}$-algebra.

It also worths being noted that the stalk $\mathcal{O}_{V_\Delta,p}$ of $\mathcal{O}_{V_\Delta}$ at $p \in V_\Delta$ identifies with $\mathcal{O}_{V,p}$ and that the quotient field $\mathcal{U}(V)$ of V identifies with $\mathcal{U}\langle\Sigma\rangle$. Consequently the functor $V \mapsto V_\Delta$ induces a functor

$$\{\text{D-varieties}\} \to \{\Delta_0\text{-manifolds}\}$$

Now the functor $V \mapsto V_\Delta$ commutes with products so it induces functors $G \mapsto G_\Delta$

$$\{\text{integral D-group schemes of D-finite type}\} \to \{\Delta\text{-groups}\}$$

$$\{\text{irreducible algebraic D-groups}\} \to \{\Delta_0\text{-groups}\}$$

A few easy remarks about the functor $V \mapsto V_\Delta$ are in order

1) If X is a D-variety then X_Δ is Zariski dense in X.

2) If $X \to Y$ is a surjective D-map of D-varieties then the map $X_\Delta \to Y_\Delta$ is surjective.

3) If X is a D-variety and $Y \subset X$ is a closed D-subvariety then $X_\Delta \cap Y = Y_\Delta$ (as subsets of X).

(2.3) It will be useful to express the functors $X \to \hat{X}$ and $X_0 \to \check{X}_0$ defined in (1.12) and (1.13) respectively in terms of the functor $V \mapsto V_\Delta$ above. Start by recalling that we introduced in (I. 2.1) a functor

$$\{\text{reduced } \mathcal{U}\text{-schemes}\} \to \{\text{reduced D-schemes}\},\quad X \mapsto X^\infty$$

which is an adjoint for the forgetful functor $V \mapsto V^!$.
Now note that if $X = \mathrm{Spec}\,\mathcal{U}[y_1,\dots,y_n]/J$ then $X^\infty \simeq \mathrm{Spec}\,\mathcal{U}\{y_1,\dots,y_n\}/\{J\}$; this can be checked by simply checking that

$$\mathrm{Hom}_{\mathcal{U}\text{-sch}}(V^!, \mathrm{Spec}\,\mathcal{U}[y_1,\dots,y_n]/J) = \mathrm{Hom}_{\text{D-sch}}(V, \mathrm{Spec}\,\mathcal{U}\{y_1,\dots,y_n\}/\{J\})$$

for any reduced D-scheme V. But by Kolchin's result quoted in (1.12) if J is prime then so is $\{J\}$. Consequently, if X is an $\mathcal{U}$-variety then X^∞ is an integral D-scheme of D-finite type so we dispose of a functor

$$\{\mathcal{U}\text{-varieties}\} \to \{\text{integral D-schemes of D-finite type}\},\quad X \mapsto X^\infty$$

We leave to the reader the task to check that the composition of the above functor with the functor in (2.2):

$$\{\text{integral D-schemes of D-finite type}\} \to \{\Delta\text{-manifolds}\},\quad V \mapsto V_\Delta$$

is isomorphic to the functor in (1.12):

$$\{\mathcal{U}\text{-varieties}\} \to \{\Delta\text{-manifolds}\},\quad X \mapsto \hat{X}$$

In other words we have a natural isomorphisms

$$\hat{X} \simeq (X^{\infty})_{\Delta}$$

(2.4) Similarly we dispose of a functor

$$\{\mathcal{K}\text{-varieties}\} \to \{D\text{-varieties}\}$$

defined by $X_o \mapsto X_o \otimes \mathcal{U}$, where $X_o \otimes \mathcal{U}$ is viewed with the split D-scheme structure (I. 3.6). It is easy to check that the composition of this functor with the functor

$$\{D\text{-varieties}\} \to \{\Delta_o\text{-manifolds}\},\quad V \mapsto V_{\Delta}$$

is isomorphic to the functor in (1.13):

$$\{\mathcal{K}\text{-varieties}\} \to \{\Delta_o\text{-manifolds}\},\quad X_o \mapsto \check{X}_o$$

In other words we have natural isomorphisms

$$\check{X}_o \simeq (X_o \otimes \mathcal{U})_{\Delta}$$

(2.5) Similarily for any irreducible algebraic $\mathcal{U}$-group G we have a natural isomorphism of Δ-groups

$$\hat{G} \simeq (G^{\infty})_{\Delta}$$

and for any irreducible algebraic $\mathcal{K}$-group G_o we have a natural isomorphism of Δ_o-groups

$$\check{G}_o \simeq (G_o \otimes \mathcal{U})_{\Delta}$$

Next we prepare ourselves to prove that the category of irreducible algebraic D-groups is equivalent to the category of Δ_o-groups. The first step is the following:

(2.6) LEMMA. Let Γ be a Δ-group. The following are equivalent:

1) Γ is a Δ_o-group.

2) $\Gamma \simeq G_{\Delta}$ for some irreducible algebraic D-group G.

Proof

2) $\Rightarrow$1) is obvious.

1) $\Rightarrow$2) Pick any affine piece $U = (\operatorname{Spec} R)_{\Delta}$ of Γ with R a Δ-finitely generated integral Δ-$\mathcal{U}$-algebra. It is a fact [B_3] that any such R whose quotient field has finite transcendence

degree over $\mathcal{U}$ contains an element $f \in R$, $f \neq 0$ such that $A := R_f$ is finitely generated as a (non-differential) $\mathcal{U}$-algebra. Performing a translation we may assume that $(\mathrm{Spec}\, A)_\Delta$ contains the unit e of Γ. So in particular the multiplication and inverse of Γ induce on $(V = \mathrm{Spec}\, A, e)$ a structure of germ of algebraic group ("groupuscule algebrique" in the sense of [U]; in fact we have rational maps $V \times V \dashrightarrow V$ and $V \dashrightarrow V$ defined at (e,e) respectively at e enjoying the usual properties of group law with e as a unit whenever composition of maps makes sense). Now there exists an algebraic group G whose germ at the identity is isomorphic to the germ (V,e) (cf. [U]; this can be seen as follows: V has in particular a normal law of composition in Weil's sense hence by Weil's theorem there is an algebraic group G and a birational map $f : V \dashrightarrow G$ such that the transported law fom V to G via f coincides with the law of G. Now one checks that f must be defined at e, f(e) = e and f commutes with multiplication and inverse arround e, so f gives an isomorphism of germs $(V,e) \simeq (G,e)$). Transporting the Δ-structure from V to G via the isomorphism $(V,e) \simeq (G,e)$ we get m commuting derivations $p_1, \ldots, p_m$ on $\mathcal{U}(G)$ which agree with comultiplication $\mathcal{U}(G) \to \mathcal{U}(G \times G)$ and antipode $\mathcal{U}(G) \to \mathcal{U}(G)$ (preserving in fact both $\mathcal{O}_{G,e}$ and its maximal ideal). By (I. 1.20) $p_i \in P(G)$ so by (I. 1.1) G becomes an algebraic D-group. We see that we have a Δ-rational map $\Gamma \dashrightarrow G_\Delta$ (i.e. a morphism of Δ-manifolds from and open subset of Γ to G_Δ) which agrees with multiplication whenever operations make sense. Exactly as in the case of algebraic groups such a map must be everywhere defined on Γ and our lemma is proved.

(2.7) LEMMA. Assume G and G' are irreducible algebraic D-groups. Then the natural map

$$\mathrm{Hom}_{\mathrm{alg.D\text{-}gr}}(G,G') \to \mathrm{Hom}_{\Delta\text{-}\mathrm{gr}}(G_\Delta, G'_\Delta)$$

is bijective.

Proof. The morphism $G_\Delta \to G'_\Delta$ induces a commutative diagram of local Δ-rings

$$\begin{array}{ccc} \mathcal{O}_{G'_\Delta,e'} = \mathcal{O}_{G',e'} & \longrightarrow & \mathcal{O}_{G,e} = \mathcal{O}_{G_\Delta,e} \\ \downarrow & & \downarrow \\ \mathcal{O}_{G'_\Delta \times G'_\Delta, e' \times e'} = \mathcal{O}_{G' \times G', e' \times e'} & \longrightarrow & \mathcal{O}_{G \times G, e \times e} = \mathcal{O}_{G_\Delta \times G_\Delta, e \times e} \end{array}$$

This extends uniquely to a commutative diagram in which the horizontal arrows are D-rational maps of D-varieties (i.e. D-maps defined on open subsets of the sourse):

$$\begin{array}{ccc} G \times G & \dashrightarrow & G' \times G' \\ \downarrow & & \downarrow \\ G & \dashrightarrow & G' \end{array}$$

The bottom horizontal arrow must then be an (everywhere defined) morphism of algebraic D-groups and our lemma is proved.

(2.8) COROLLARY. The functor $G \mapsto G_\Delta$

$$\{\text{irreducible algebraic D-groups}\} \to \{\Delta_0\text{-groups}\}$$

is an equivalence of categories.

Proof. It is just (2.6) + (2.7).

REMARK. We shall fix from now on a quasi-inverse $\Gamma \mapsto G(\Gamma)$ of the functor in (2.8). We have

$$\mathrm{Hom}_{\Delta\text{-gr}}(\Gamma,\Gamma') \simeq \mathrm{Hom}_{\mathrm{alg\,D\text{-}gr}}(G(\Gamma),G(\Gamma'))$$

Moreover $\mathcal{U}\langle\Gamma\rangle$ identifies with $\mathcal{U}\langle G(\Gamma)\rangle$ and $\mathcal{O}_{\Gamma,e}$ identifies with $\mathcal{O}_{G(\Gamma),e}$.

(2.9) Clearly a Δ_0-group Γ is split if and only if $G(\Gamma)$ is a split algebraic D-group. But by (I. 3.7) the latter happens if and only if $G(\Gamma)$ is locally finite. So by (IV, 1.2) upon letting $G = G(\Gamma)^!$ we have

$$\Gamma \text{ is split iff the images of } \delta_1,\dots,\delta_m \in \Delta \subset P \text{ in } P(G) \text{ belong to } P(G,\mathrm{fin})$$

(recall that any algebraic D-group structure on G is determined by a Lie $\mathcal{U}/\mathcal{K}$-algebra map $P \to P(G)$, cf. (I. 1.1)).

(2.10) LEMMA. For any Δ_0-group Γ and any irreducible algebric $\mathcal{U}$-group H we have a natural bijection

$$\mathrm{Hom}_{\Delta\text{-gr}}(\Gamma,\hat{H}) \simeq \mathrm{Hom}_{\mathrm{alg}\,\mathcal{U}\text{-gr}}(G(\Gamma)^!,H)$$

Proof. By (I. 2.1) we have

$$\mathrm{Hom}_{\mathrm{alg}\,\mathcal{U}\text{-gr}}(G(\Gamma)^!,H) \simeq \mathrm{Hom}_{\mathrm{D\text{-}gr\,sch}}(G(\Gamma),H^{\infty})$$

Since $\Gamma = G(\Gamma)_{\Delta}$ there is an obvious map

$$\mathrm{Hom}_{\mathrm{D\text{-}gr\,sch}}(G(\Gamma),H^{\infty}) \to \mathrm{Hom}_{\Delta\text{-gr}}(\Gamma,\hat{H})$$

Let's construct an inverse to it; start with $f \in \mathrm{Hom}_{\Delta\text{-gr}}(\Gamma,\hat{H})$. Then f provides a Δ-ring map

$$\mathcal{O}_{H^{\infty!},e} = \mathcal{O}_{H^{\infty},e} = \mathcal{O}_{\hat{H},e} \to \mathcal{O}_{\Gamma,e} = \mathcal{O}_{G(\Gamma),e}$$

Composing this map with the natural map

$$\mathcal{O}_{H,e} \to \mathcal{O}_{H^{\infty!},e}$$

we get a map

$$\mathcal{O}_{H,e} \to \mathcal{O}_{G(\Gamma),e}$$

hence a rational map $G(\Gamma)^! \to H$ which is easily seen to agree with multiplication generically and hence is an (everywhere defined) homomorphism. This homomorphism provides a morphism $G(\Gamma) \to H^\infty$ and our inverse is constructed.

(2.11) LEMMA. For any irreducible algebraic $\mathcal{K}$-groups G_o and H_o we have a natural bijection

$$\mathrm{Hom}_{\mathrm{alg}\,\mathcal{K}\text{-gr}}(G_o,H_o) \simeq \mathrm{Hom}_{\Delta\text{-gr}}(\check{G}_o,\check{H}_o)$$

Proof. By (I. 3.6) we have a bijection

$$\mathrm{Hom}_{\mathrm{alg}\,\mathcal{K}\text{-gr}}(G_o,H_o) \simeq \mathrm{Hom}_{\mathrm{alg}\,D\text{-gr}}(G_o \otimes \mathcal{U}, H_o \otimes \mathcal{U})$$

and we conclude by (2.7).

(2.12) One can define in the obvious way actions of Δ-groups on Δ-manifolds (and on Δ-groups) so one can define Δ-cocycles of a Δ-group Γ in a Δ-group Γ' on which Γ acts.

Some more definitions: a morphism $\Gamma \to \Gamma'$ of Δ-groups will be called an embedding if it is injective. A Δ-group Γ will be called linear (respectively abelian) if there is an embedding $\Gamma \to \widehat{GL}_N$ for some $N \geq 1$ (respectively $\Gamma \to \hat{A}$ for some abelian $\mathcal{U}$-variety $\hat{A}$). A Δ-group Γ will be said to have no non-trivial linear representation if for any $N \geq 1$ the only morphism $\Gamma \to \widehat{GL}_N$ is the trivial one. By (2.10) Γ has no non-trivial linear representation iff $G(\Gamma)$ has the same propety i.e. iff $\mathcal{O}(G(\Gamma)) = \mathcal{U}$. Let Γ be a Δ-group. Two embeddings $\alpha' : \Gamma' \to \Gamma$, $\alpha'' : \Gamma'' \to \Gamma$ are called equivalent if there is an isomorphism $\sigma : \Gamma' \to \Gamma''$ such that $\alpha''\sigma = \alpha'$. A class of equivalence of such embeddings will be called a Δ-subgroup of Γ. By abuse we will sometimes say (given an embedding $\Gamma' \to \Gamma$) that Γ' is a Δ-subgroup of Γ.

(2.13) PROPOSITION. Let Γ be a Δ_o-group and let $G = G(\Gamma)^!$. Then there is an embedding $\Gamma \to \hat{G}$ and a Δ-cocycle $\hat{G} \to (L(G)^m)^\wedge$ inducing an exact sequence of pointed sets

$$1 \to \Gamma \to G(\mathcal{U}) \to L(G)^{\mathrm{int}} \to 1$$

where the action of $\hat{G}$ on $(L(G)^m)^\wedge = (L(G)^\wedge)^m$ is induced by adjoint action ($L(G)$ is viewed as usual as an algebraic vector group) and

$$L(G)^{\mathrm{int}} = \{(\theta_1,\ldots,\theta_m) \in L(G)^m;\ \nabla_{\delta_i}\theta_j - \nabla_{\delta_j}\theta_i + [\theta_i,\theta_j] = 0 \text{ for all } i,j\}$$

where ∇ here is the adjoint connection (I. 1.10).

Proof. It's just a translation of (I. 2.3) via the yoga of the present section.

REMARK. The above proposition answers (in the finite dimensional case) Kochin's question from the Introduction of $[K_2]$ whether differential algebraic groups can be embedded into algebraic groups.

(2.14) In what follows by a Hopf Δ-$\mathcal{U}$-algebra we mean a Hopf $\mathcal{U}$-algebra H which is a Δ-$\mathcal{U}$-algebra such that comultiplication, antipode and unit are Δ-algebra maps; this concept is equivalent of course to that of Hopf D-algebra. There is a functor induced by that in (1.2):

$$\{\text{integral } \Delta\text{-finitely generated commutative Hopf } \Delta\text{-}\mathcal{U}\text{-algebras}\} \to \{\text{affine } \Delta\text{-groups}\}$$
$$H \mapsto (\operatorname{Spec} H)_\Delta$$

Any Δ-group isomorphic to an object contained in the image of the above functor will be called strictly affine. It can be shown (but won't be done or used here) that our strictly affine Δ-groups "are" precisely Cassidy's Δ-algebraic groups with Δ-polynomial law $[C_1]$ and consequently by Cassidy's results $[C_1]$ any strictly affine Δ-group is linear. The latter result for Δ_o-groups will be reproved below.

The following result is due to Cassidy $[C_3]$ but we provide for it a direct proof:

(2.15) LEMMA. For any two integral Δ-finitely generated commutative Hopf Δ-$\mathcal{U}$-algebras H and H' there is a bijection:

$$\operatorname{Hom}_{\text{Hopf } \Delta\text{-}\mathcal{U}\text{-alg}}(H',H) \simeq \operatorname{Hom}_{\Delta\text{-gr}}((\operatorname{Spec} H)_\Delta, (\operatorname{Spec} H')_\Delta)$$

Proof. The natural map from left to right is clearly injective. To check surjectivity take a morphism $f : (\operatorname{Spec} H)_\Delta \to (\operatorname{Spec} H')_\Delta$; it induces a morphism

$$\mathcal{O}_{\operatorname{Spec} H',e} = \mathcal{O}_{(\operatorname{Spec} H')_\Delta,e} \to \mathcal{O}_{(\operatorname{Spec} H)_\Delta,e} = \mathcal{O}_{\operatorname{Spec} H,e}$$

Writing both H and H' as direct limits of finitely generated Hopf algebras [DG] one immediately sees that the latter morphism is induced by a morphism of affine group schemes $\operatorname{Spec} H \to \operatorname{Spec} H'$ hence by a Hopf Δ-$\mathcal{U}$-algebra map $H' \to H$ and we are done.

(2.16) Due to (2.15) we may define correctly (up to an isomorphism) for any strictly affine Δ-group Γ a Δ-finitely generated Hopf Δ-$\mathcal{U}$-algebra $\mathcal{U}\{\Gamma\}$ such that $\Gamma \simeq (\operatorname{Spec}\mathcal{U}\{\Gamma\})_\Delta$; this $\mathcal{U}\{\Gamma\}$ can be called the Δ-coordinate Hopf algebra of Γ (and coincides with Cassidy's $\mathcal{A}\{\Gamma\}$ in $[C_1]$!). It does not coincide however with $\mathcal{O}(\Gamma)$ in general!

(2.17) LEMMA $[B_3]$. Let Γ be a strictly affine Δ_o-group. Then $\mathcal{U}\{\Gamma\}$ is finitely generated as a (non-differential) $\mathcal{U}$-algebra. In particular $G(\Gamma) = \operatorname{Spec}\mathcal{U}\{\Gamma\}$ and $\mathcal{U}\{\Gamma\} = \mathcal{O}(G(\Gamma))$. Moreover Γ is split if and only if $\mathcal{U}\{\Gamma\}$ is a locally finite D-module.

Proof. We already used in (2.6) the fact (proved in $[B_3]$) that if $\Gamma = (\operatorname{Spec} H)_\Delta$ then the scheme Spec H contains a non-empty open subset of finite type over $\mathcal{U}$. A translation argument shows that one can cover Spec H by affine open sets of finite type over $\mathcal{U}$, hence Spec H is of finitey type over $\mathcal{U}$, hence H is finitely generated over $\mathcal{U}$. The remaining assertions follow immediately from (2.8) and (2.10).

(2.18) COROLLARY. A Δ_o-group Γ is strictly affine if and only if $G(\Gamma)^!$ is affine, if and only if Γ is linear.

Proof. The first equivalence follows from (2.17). The second one follows from (2.17), (2.13) and (2.10).

(2.19) REMARK. By (1.25) we see that there exist affine Δ_o-groups Γ which are not strictly affine (e.g. $\Gamma = \check{A}_o$ with A_o an abelian $\mathcal{K}$-variety; this example was first discovered by Cassidy).

Let's analyse a few

EXAMPLES

It worths describing explicitely the algebraic D-groups associated to the Δ_o-groups introduced in Examples 1), 2), 3) given after (1.12).

1) Assume first that $\Sigma = Z(y'' + ay' + by) \subset \hat{G}_a = \mathcal{U}$, $a,b \in \mathcal{U}$. Then $G(\Sigma) = G_a^2 = \mathrm{Spec}\,\mathcal{U}[\xi_1,\xi_2]$ viewed as an algebraic D-group via the derivation $\tilde{\delta} \in P(G_a^2)$ defined by $\tilde{\delta}\xi_1 = \xi_2$ and $\tilde{\delta}\xi_2 = -b\xi_1 - a\xi_2$ i.e. by

$$\tilde{\delta} = \delta^* + \xi_2 \frac{\partial}{\partial \xi_1} - (b\xi_1 + a\xi_2)\frac{\partial}{\partial \xi_2}$$

where δ^* is the lifting of δ to $\mathcal{U}[\xi_1,\xi_2]$ which kills ξ_1,ξ_2. This structure is of course a special case of the corresponding example in (1.9), (1.26). To check that G_a^2 viewed as an algebraic D-group as above is indeed isomorphic to Σ we must check that $(G_a^2)_\Delta \simeq \Sigma$. But $(G_a^2)_\Delta$ is by definition the set of all $\mathcal{U}$-points $p = (\alpha_1,\alpha_2) \in G_a^2$ for which the evaluation map $\mathcal{U}[\xi_1,\xi_2] \to \mathcal{U}$ is a D-algebra map hence for which

$$\alpha_1' = \alpha_2$$

$$\alpha_2' = -b\alpha_1 - a\alpha_2$$

i.e. for which

$$\alpha_1' = \alpha_2$$

$$\alpha_1'' + a\alpha_1' + b\alpha_1 = 0$$

We checked that $(G_a^2)_\Delta \simeq \Sigma$ as sets; the reader may check that this is an isomorphism of Δ-manifolds. Note that $\mathcal{U}\langle\Sigma\rangle = \mathcal{U}(y,y')$ where y is the class of $y \in \mathcal{U}\{y\}$ mod$[y'' + ay' + by]$. Under the identification $\mathcal{U}(y,y') = \mathcal{U}(\xi_1,\xi_2)$ we have $y = \xi_1$, $y' = \xi_2$.

2) Similarily if we assume $\Sigma = Z(xy - 1, yy'' - (y')^2 + ayy') \subset \hat{G}_m = Z(xy - 1) \subset \mathcal{U}^2$ as in (1.12) then one can check as above that $G(\Sigma) = G_m \times G_a = \mathrm{Spec}\,[\chi,\chi^{-1}\xi]$ as in (1.9), (1.26)

iewed as an algebraic D-group via the derivation $\tilde{\delta} \in P(G_m \times G_a)$ defined by $\tilde{\delta}\chi = \xi\chi$ and $\tilde{\delta}\xi = -a\xi$

i.e. by

$$\tilde{\delta} = \delta^* + \xi\chi\frac{\partial}{\partial\chi} - a\xi\frac{\partial}{\partial\xi}$$

Once again we have $\mathcal{U}\langle\Sigma\rangle = \mathcal{U}(y,y')$. Under the identification $\mathcal{U}(y,y') = \mathcal{U}(\chi,\xi)$ we have $\chi = y$ and $\xi = y'/y$.

3a) Finally if we let Γ be the Δ-closure in $\hat{\mathbf{P}}^2$ of $Z(y^2 - x(x-1)(x-c), x''y - x'y' + ax'y) \subset \mathcal{U}^2 = \mathbf{A}^2$ then one can check as above that $G(\Gamma) = A \times G_a = A \times \operatorname{Spec}\mathcal{U}[\xi]$ as in (1.9) (with A = Zariski closure of $Z(y^2 - x(x-1)(x-c)) \subset \mathcal{U}^2$ in $\mathbf{P}^2$) viewed as an algebraic D-group via the derivation $\tilde{\delta} \in P(A \times G_a)$ defined by $\tilde{\delta} x = \xi y$ and $\tilde{\delta}\xi = -a\xi$ hence by

$$\tilde{\delta} = \delta^* + \xi y\frac{\partial}{\partial x} - a\xi\frac{\partial}{\partial\xi}$$

We have $\mathcal{U}\langle\Gamma\rangle = \mathcal{U}(x,y,x')$; under the identification $\mathcal{U}(x,y,x') = \mathcal{U}(x,y,\xi)$ we have $\xi = x'/y$.

Note that in all three examples our Δ_o-groups appear embedded in 1-dimensional algebraic $\mathcal{U}$-groups (more precisely in $\hat{G}_a, \hat{G}_m, \hat{A}$) and are given by second order differential equations. These embeddings are entirely different from the embeddings provided by (2.13) above (which should be viewed as the "natural" embeddings, in which the ambient Δ-group should be $\hat{G}_a \times \hat{G}_a$, $\hat{G}_m \times \hat{G}_a$, $\hat{A} \times \hat{G}_a$). Concretely, in example 1) Σ is given in $\hat{G}_a \times \hat{G}_a$ by the first order system

$$\xi_1' - \xi_2 = 0$$
$$\xi_2' + b\xi_1 + a\xi_2 = 0$$

In example 2) Σ is given in $\hat{G}_m \times \hat{G}_a$ by the first order system

$$\chi' - \xi\chi = 0$$
$$\xi' + a\xi = 0$$

In example 3) Γ is given in $\hat{A} \times \hat{G}_a$ by the first order system

$$x' - \xi y = 0$$
$$\xi' + a\xi = 0$$

We are not yet ready to compute $G(\Gamma)$ for Γ as in Example 3b) of (1.12). See (3.20) for a treatment of it. But we are ready to compute the Δ-coordinate Hopf algebra $R = \mathcal{U}\{\Sigma\}$ of the Δ-group

$$\Sigma = \left\{ \begin{pmatrix} a & 0 & b \\ 0 & 1 & c \\ 0 & 0 & 1 \end{pmatrix} \in GL_3(\mathcal{U});\ c' = 0,\ a' = ca \right\}$$

in Example 4) of loc.cit. Indeed, as a Δ-algebra

$$R = \mathcal{U}\{a,b,c\}_a/[c',a' - ca] = \mathcal{U}[a,a^{-1},c,b,b',b'',b''',\ldots]$$

The computation

$$\begin{pmatrix} a_1 & 0 & b_1 \\ 0 & 1 & c_1 \\ 0 & 0 & 1 \end{pmatrix}\begin{pmatrix} a_2 & 0 & b_2 \\ 0 & 1 & c_2 \\ 0 & 0 & 1 \end{pmatrix} = \begin{pmatrix} a_1a_2 & 0 & a_1b_2 + b_1 \\ 0 & 1 & c_1 + c_2 \\ 0 & 0 & 1 \end{pmatrix}$$

gives the following formulae for the comultiplication $\mu : R \to R \otimes R$:

$$\mu a = a \otimes a$$

$$\mu c = c \otimes 1 + 1 \otimes c$$

$$\mu b = a \otimes b + b \otimes 1$$

$$\mu(b') = ca \otimes b + a \otimes b' + b' \otimes 1$$

$$\mu(b'') = c^2 a \otimes b + 2ca \otimes b' + a \otimes b'' + b'' \otimes 1$$

. . . .

Let $R_n = \mathcal{U}[a,a^{-1},c,b,b',\ldots,b^{(n)}]$; it is a Hopf subalgebra of R so we may consider the affine algebraic group $G_n = \operatorname{Spec} R_n$ and the affine group scheme $G = \operatorname{Spec} R$. Put $T = \mathcal{U}[a,a^{-1}]$. Then each G_n contains T and has a projection onto T. Moreover $\ker(G_{n+1} \to G_n)$ is a vector group for all n. Consequently each G_n is solvable hence a semidirect product of T by the unipotent group

$$M_n = \operatorname{Ker}(G_n \to T) = \operatorname{Spec}\mathcal{U}[c,b,b',\ldots,b^{(n)}]$$

So G itself is a semidirect product $M \times_\rho T$, $M = \varprojlim M_n$. Now note the remarkable fact that $a \in X_m(T)$ is a weight for ρ but is not killed by δ; this is the example promissed in (II.3.6)!

3. Structure of Δ_o-groups

Since we proposed ourselves not to use Kolchin's powerful theory $[K_2]$ we are faced with the problem of constructing a "Lie correspondence" and a theory of subgroups and quotients for Δ-groups. We are able to do this only if we restrict ourselves to Δ_o-groups. We begin with sketching such a theory.

(3.1) Let $\Gamma' \to \Gamma$ be an embedding of Δ_o-groups. We shall define a Δ-manifold structure on the set Γ/Γ' as follows. By (2.2) $G(\Gamma')(\mathcal{U}) \to G(\Gamma)(\mathcal{U})$ is injective so $G(\Gamma') \to G(\Gamma)$ is a closed immersion hence by (I. 1.18) $G(\Gamma)/G(\Gamma')$ has a natural structure of D-variety. One checks using (2.2) that the underlying set of the Δ_o-manifold

$$(G(\Gamma)/(G(\Gamma'))_\Delta$$

coincides with Γ/Γ'. So we give Γ/Γ' the above Δ_o-manifold structure.

If in addition Γ' is normal in Γ then by a density argument (cf. (2.2)) we get that $G(\Gamma')(\mathcal{U})$ is normal in $G(\Gamma)(\mathcal{U})$ hence $G(\Gamma')$ is normal in $G(\Gamma)$ hence $G(\Gamma)/G(\Gamma')$ will be an algebraic D-group so Γ/Γ' will be a Δ_o-group. The reader can check that in fact the above Γ/Γ' satisfies the usual universal property of homogenous spaces (respectively quotient groups).

(3.2) Clearly (I. 1.16) provides the desired "Lie correspondence" for Δ_o-groups; we leave its formulation to the reader.

(3.3) Let's discuss commutators. By (I. 1.6) if Γ is a Δ_o-group then $[G(\Gamma),G(\Gamma)]$ is an irreducible algebraic D-subgroup of $G(\Gamma)$. Then using (2.2) one checks that the underlying groups of the Δ_o-group

$$[G(\Gamma),G(\Gamma)]_\Delta$$

coincides with the closure of $[\Gamma,\Gamma]$ in Γ. So we denote the above Δ_o-group by $[\Gamma,\Gamma]_{cl}$.

(3.4) Let's discuss centers. By (I. 1.6) if Γ is a Δ_o-group then $Z^o(G(\Gamma))$ is an irreducible algebraic D-subgroup of $G(\Gamma)$. Using (2.2) one checks that the underlying group of the Δ_o-group

$$Z^o(G(\Gamma))_\Delta$$

coincides with the connected component $Z^o(\Gamma)$ of the center of Γ. So we give $Z^o(\Gamma)$ the above Δ_o-group structure.

(3.5) Some words about isogenies. A morphism of Δ_o-groups $\Gamma \to \Gamma'$ will be called an isogeny if it is surjective with finite kernel. By a density argument again cf. (2.2) the homomorphism $G(\Gamma)(\mathcal{U}) \to G(\Gamma')(\mathcal{U})$ is surjective; moreover, if H is the identity component of $Ker(G(\Gamma) \to G(\Gamma'))$ by (2.2) we have $H_\Delta = H(\mathcal{U}) \cap \Gamma$ is dense in $H(\mathcal{U})$, this showing that $H(\mathcal{U})$ consists of one element only. Hence $G(\Gamma) \to G(\Gamma')$ must also be an isogeny.

Here is one of the main applications:

(3.6) THEOREM. Let Γ be Δ_o-group. Put $G = G(\Gamma)$ and identify $\mathcal{U}\langle\Gamma\rangle$ with $\mathcal{U}\langle G\rangle$. Then

1) Γ is split if and only if $\Gamma/[\Gamma,\Gamma]$ is split.

2) Γ is split if and only if the maximum semiabelian subfield of $\mathcal{U}\langle\Gamma\rangle$ is a Δ-subfield of $\mathcal{U}\langle\Gamma\rangle$.

3) If $\Gamma \to \Gamma'$ is an isogeny then Γ is split if and only if Γ' is split.

4) If $Z^o(\Gamma) = \{e\}$ then Γ is split.

Proof. 1), 2) follow from (2.9), (3.3) and (IV. 1.8).

3) follows from (2.9), (3.5) and (IV. 1.8).

4) follows from (2.9), (3.4) and (II. 3.30).

(3.7) A linear Δ-group Γ will be called solvable (respectively nilpotent) if it is so as an abstract group.

Note that a linear Δ_0-group Γ is solvable (respectively nilpotent) if and only if $G = G(\Gamma)^!$ is so (this follows from "density", cf. (2.2) and from (2.13)).

A Δ-group will be called unipotent if there exists an embedding $\Gamma \to \hat{GL}_N$ whose image consists of unipotent matrices (this definition agrees with Cassidy's $[C_2]$). Note that a Δ_0-group Γ is unipotent if and only if $G(\Gamma)$ is so; indeed if $G(\Gamma)$ is unipotent so is Γ by (2.13). Conversely if Γ embeds into $\hat{GL}_N$ and Γ consists of unipotent matrices of $GL_N(\mathcal{U})$ then one can easily show that the action of Γ via translations on the algebra $\mathcal{U}\{GL_N\} = \mathcal{O}(GL_N^\infty)$ is locally unipotent hence so will be the action of Γ on $\mathcal{U}\{\Gamma\} = \mathcal{O}(G(\Gamma))$ hence so will be the action of $G(\Gamma)(\mathcal{U})$ on $\mathcal{O}(G(\Gamma))$ (by "density" again) and we are done (see [H] p. 64 for background of locally unipotent actions).

Let Γ be a linear Δ_0-group. By (2.18) $G(\Gamma)$ is affine and by (II. 3.30) its radical $R(G(\Gamma))$ is an irreducible algebraic D-subgroup of $G(\Gamma)$. We define the radical of Γ by the formula

$$R(\Gamma) := R(G(\Gamma))_\Delta$$

We have an embedding $R(\Gamma) \to \Gamma$ and it is easy to check using (2.8) that for any embedding $\Gamma' \to \Gamma$ of linear Δ_0-groups with Γ' solvable and normal in Γ there is a commutative diagram

$$\begin{array}{ccc} \Gamma' & \longrightarrow & \Gamma \\ & {\scriptstyle\alpha}\searrow & \nearrow \\ & R(\Gamma) & \end{array}$$

with α unique. Here is another consequence of our theory.

(3.8) THEOREM. A linear Δ_0-group Γ is split if and only if its radical $R(\Gamma)$ is split, if and only if all group like elements of the Hopf Δ-$\mathcal{U}$-algebra $\mathcal{U}\{\Gamma\}$ are Δ-constants.

Proof. It follows directly from (3.7), (2.9), (II. 3.30).

In what follows we briefly discuss abelian Δ_0-groups:

(3.9) LEMMA. Let G be an irreducible algebraic D-group, H an irreducible algebraic $\mathcal{U}$-group and $G \to H^\infty$, $G^! \to H$ two morphisms which correspond to eachother under adjunction (I. 2.1). The following are equivalent:

1) The kernel of $G \to H^\infty$ is trivial.

2) The kernel of $G^! \to H$ contains no non-trivial algebraic D-subgroup of G.

3) The morphism $G_\Delta \to \hat{H}$ is an embedding.

Proof. 1)$\Rightarrow$3) is clear.

2)$\Rightarrow$1) follows since $\mathrm{Ker}(G \to H^\infty)$ is an algebraic D-subgroup of G contained in $\mathrm{Ker}(G^! \to H)$.

1)$\Rightarrow$2) Assume M is an algebraic D-subgroup of G contained in $\mathrm{Ker}(G^! \to H)$. Then both the trivial D-morphism

$$\phi : M \to \mathrm{Spec}\,\mathcal{U} \xrightarrow{\varepsilon} H^{\infty}$$

and the D-morphism

$$\psi : M \to G \to H^{\infty}$$

composed with $(H^{\infty})^! \to H$ give the trivial morphism. By universality, $\phi = \psi$ so M is trivial.

(3.10) LEMMA. Let Γ be a Δ_0-group. The following are equivalent:

1) Γ is abelian.

2) There is a D-morphism $G(\Gamma) \to A^{\infty}$ with trivial kernel with A an abelian variety.

3) The linear part of $G(\Gamma)^!$ contains no non-trivial algebraic D-subgroup of $G(\Gamma)$.

4) Any morphism from a linear Δ_0-group Γ' to Γ is trivial.

Moreover, if the above conditions hold then the linear part of $G(\Gamma)^!$ is unipotent.

Proof. Let B be the linear part of $G^!$ $(G = G(\Gamma))$.

3)$\Rightarrow$2) follows from (3.9) applied to the projection $G^! \to G^!/B$.

2)$\Rightarrow$1) follows from (2.10) and (3.9).

2)$\Rightarrow$3). We have a commutative diagram

$$\begin{array}{ccc} G^! & \longrightarrow & A \\ & \searrow \quad \nearrow & \\ & G^!/B & \end{array}$$

providing a commutative diagram

$$\begin{array}{ccc} G & \xrightarrow{\ \phi\ } & A^{\infty} \\ \psi \searrow & & \nearrow \\ & (G^!/B)^{\infty} & \end{array}$$

since ϕ has a trivial kernel, so has ψ. Applying (3.9) to ψ we get our conclusion. Note that if 2) or 3) hold G is commutative so by (I. 1.22) the maximal torus B_m of B is an algebraic D-subgroup of G hence is trivial so B is unipotent.

3)$\Rightarrow$4) If $\Gamma' \to \Gamma$ is as in 4) then the image of $G(\Gamma') \to G(\Gamma)$ is an algebraic D-subgroup of B hence is trivial. So $G(\Gamma') \to G(\Gamma)$ is trivial so $\Gamma' \to \Gamma$ is trivial.

4)$\Rightarrow$3) Since $[\Gamma,\Gamma]_{cl}$ is linear it is trivial so Γ is commutative. By (I. 1.22) once again B_m is an algebraic D-subgroup of G. By 4) B_m is trivial. Now assume 3) does not hold hence there exists an algebraic D-subgroup H of G contained in B with $H \neq 0$. Since H is unipotent it is irreducible. Letting $\Gamma' = H_{\Delta}$ we get a contradiction. Our lemma is proved.

(3.11) COROLLARY. 1) Any Δ_o-group Γ possess a linear Δ_o-subgroup Γ' such that Γ/Γ' is an abelian Δ_o-group.

2) Any Δ_o-group Γ possesses a Δ_o-subgroup Γ_1 with no non-trivial linear representation such that Γ/Γ_1 is a linear Δ_o-group.

Proof. 1) Take $\Gamma' = G'_\Delta$ where G' is the largest irreducible affine algebraic D-subgroup of $G(\Gamma)$ and use (3.10).

2) Take $\Gamma_1 = (G_1)_\Delta$ where $G_1 = \mathrm{Ker}(G \to \mathrm{Spec}\,\mathcal{O}(G))$ and use [DG] p. 358.

(3.12) REMARK. The first of the above assertions answers, in the particular case of Δ_o-groups, Kolchin's question in the Introduction to $[K_2]$ whether a "Chevalley structure theorem" [Ro] holds for differential algebraic groups. In the case of Δ_o-groups one sees that the role of abelian varieties among algebraic groups is played by abelian Δ_o-groups. This justifies a special study of the latter. Note that we have the following inclusions of sets

$$\mathfrak{M}^{ab} = \left\{\begin{array}{l}\text{abelian } \Delta_o\text{-groups}\\ \text{modulo isomorphisms}\end{array}\right\} \hookrightarrow \mathfrak{M}^{comm} = \left\{\begin{array}{l}\text{commutative } \Delta_o\text{-groups}\\ \text{modulo isomorphism}\end{array}\right\} \hookleftarrow \mathfrak{M}^{\#} = \left\{\begin{array}{l}\Delta_o\text{-groups with non-trivial linear}\\ \text{representation modulo isomorphism}\end{array}\right\}$$

It is easy to show that $\mathfrak{M}^{ab} \not\subset \mathfrak{M}^{\#}$ (take the preimage via logarithmic derivative $A(\mathcal{U}) \to L(A)$, where A is an abelian $\mathcal{U}$-variety defined over $\mathcal{K}$, of a Δ_o-subgroup of L(A)) and that $\mathfrak{M}^{\#} \not\subset \mathfrak{M}^{ab}$ (take a split Δ_o-group Γ such that $\mathcal{O}(G(\Gamma)) = \mathcal{U}$ and $G(\Gamma)$ is not an abelian variety). In what follows we concentrate on objects of the set

$$\mathfrak{M}^{ab,\#} := \mathfrak{M}^{ab} \cap \mathfrak{M}^{\#}$$

of isomorphism classes of abelian Δ_o-groups with no non-trivial linear representation.

(3.13) LEMMA. Let Γ be a Δ_o-subgroup of $\hat{H}$ with H an irreducible algebraic $\mathcal{U}$-group. The following are equivalent:

1) Γ is Zariski dense in $\hat{H}$

2) The morphism $G(\Gamma)^! \to H$ (cf. (2.10)) is surjective.

Proof. Use definitions.

(3.14) LEMMA. Let A be an abelian $\mathcal{U}$-variety and $\Gamma \to \hat{A}$ a morphism from a Δ_o-group Γ. The following are equivalent:

1) Γ is a Zariski dense Δ-subgroup of $\hat{A}$.

2) Γ is a Δ-subgroup of $\hat{A}$ containing the torsion subgroup of A.

3) The morphism $G(\Gamma)^! \to A$ (cf. (2.10)) is surjective, its kernel coincides with the linear

part B of $G(\Gamma)^!$ and B contains no non-trivial algebraic D-subgroup of $G(\Gamma)$.

(3.15) COROLLARY. Any abelian Δ_o-group Γ has a Zariski dense embedding $\Gamma \to \hat{A}$ into some produced abelian variety $\hat{A}$ Moreover for any two such embeddings $\alpha_1 : \Gamma \to \hat{A}_1, \alpha_2 : \Gamma \to \hat{A}_2$ there exists an isomorphism $\sigma : A_1 \to A_2$ such that $\hat{\sigma} \circ \alpha_1 = \alpha_2$.

Proof of (3.14). In view of (3.9), (3.10) and (3.13) in order to check that 1)$\Leftrightarrow$3) it is sufficient to check that under the assumption that $G(\Gamma)^! \to A$ is surjective and that its kernel M contains no non-trivial algebraic D-subgroup of $G(\Gamma)$ it follows that M equals the linear part B of $G(\Gamma)$. Indeed by (I. 1.22) M^o must be linear hence $M^o = B$ (actually by loc.cit. B must be an algebraic vector group). We want to prove $M = B$; assume $M/B \neq 0$ and look for a contradiction. Take $x \in M$, $x \notin B$, let $N = |M/B|$ hence $Nx \in B$; choose (B being divisible!) $b \in B$ such that $Nb = Nx$. Then $x - b \in \mathrm{Tors}(G(\Gamma))$ hence by (I. 1.22) again the (finite) subgroup of $G(\Gamma)$ generated by $x - b$ is a non-trivial algebraic D-subgroup of $G(\Gamma)$ contained in M, contradiction. So the equivalence 1)$\Leftrightarrow$3) is proved.

The implication 2)$\Rightarrow$1) is obvious. Conversely to check that 1)$\Rightarrow$2) it is sufficient to prove that any torsion point $x \in A(\mathcal{U})$ lifts to a D-point of $G(\Gamma)$. But any x as above lifts to a torsion point of $G(\Gamma)$ which is automatically a D-point due to (I. 1.22) again.

(3.16) Let Γ be an abelian Δ_o-group. By its embedding dimension we will understand the dimension of any abelian $\mathcal{U}$-variety A such that there exists a Zariski *dense embedding* $\Gamma \to \hat{A}$. Clearly the embedding dimension of Γ equals the dimension of the abelian part of $G(\Gamma)^!$ (cf. (3.14), (3.15)).

(3.17) Recall from (III. 1.1), (III. 1.3) cf. [KO] that for any abelian $\mathcal{U}$-variety A, $H^1_{DR}(A)$ has a D-module structure induced by Gauss-Manin connection. Call $DH^o(\Omega^1_{A/\mathcal{U}})$ the D-submodule of $H^1_{DR}(A)$ generated by $H^o(\Omega^1_{A/\mathcal{U}})$. Moreover recall from (III. 2.15) that the universal extension E(A) has a (unique) structure of algebraic D-group hence we dispose of a naturally associated Δ_o-group $E(A)_\Delta$, cf. (1.9). Here is one of our main applications of the theory in (III. 2):

(3.18) THEOREM. Let A be an abelian $\mathcal{U}$-variety. Then $\hat{A}$ contains a unique Zariski dense Δ_o-subgroup with no non-trivial linear representation (call it $A^{\#}$) and we have the formula

$$\dim A^{\#} = \dim_{\mathcal{U}} (DH^o(\Omega^1_{A/\mathcal{U}}))$$

Moreover $A^{\#}$ is a quotient of $E(A)_\Delta$.

Proof. Let Γ be any Zariski dense Δ_o-subgroup of $\hat{A}$ Such Γ's correspond bijectively to equivalence classes of surjections $G^! \to A$ (G an irreducible algebraic D-group) whose kernel is the linear part of $G^!$ and contains no non-trivial algebraic D-subgroups of G, cf. (3.14), (2.8); two such surjections $G_1^! \to A$, $G_2^! \to A$ are equivalent if there is an algebraic D-group isomorphism $\sigma : G_1 \to G_2$ making the following diagram commutative:

(3.18.1)
$$\begin{array}{ccc} G_1^! & & \\ {\scriptstyle \sigma^!}\downarrow & \searrow & \\ & & A \\ & \nearrow & \\ G_2^! & & \end{array}$$

Now assume in addition that Γ has no non-trivial linear representation (by (2.10) this is equivalent to $G(\Gamma)^!$ having the same property). By (3.10) and (III. 2.5) $G(\Gamma)^!$ is an extension of A by a vector group B such that the map $X_a(B) \to H^1(\mathcal{O}_A)$ is injective. So $G(\Gamma)^!$ is isomorphic over A with a quotient of the universal extension E(A) be some linear group. By (III. 2.2) the surjection $E(A) \to G(\Gamma)$ is a D-morphism. In view of our equivalence relation (3.18.1) we may assume $G(\Gamma) = E(A)/H$ for some linear algebraic D-subgroup H of E(A). We claim that $S_{DR}(E(A)/H)_a = DH^0(\Omega^1_{A/\mathcal{U}})$; indeed $S_{DR}(E(A)/H)_a$ contains $H^0(\Omega^1_{A/\mathcal{U}})$ and since it is a D-submodule of $H^1_{DR}(A)$, cf. (III. 2.2), it must contain $DH^0(\Omega^1_{A/\mathcal{U}})$. Assume $DH^0(\Omega^1_{A/\mathcal{U}})$ is strictly contained in $S_{DR}(E(A)/H)_a$. Then by (III. 2.14) we have

$$L(H) = S_{DR}(E(A)/H)_a^{\perp} \subset DH^0(\Omega^1_{A/\mathcal{U}})^{\perp} \subset H^0(\Omega^1_{A/\mathcal{U}})^{\perp} = S_{DR}(A)_a^{\perp} = L(\tilde{B})$$

where $\tilde{B} = \mathrm{Ker}(E(A) \to A)$. Now $DH^0(\Omega^1_{A/\mathcal{U}})^{\perp} = L(H_1)$ for some intermediate algebraic vector group H_1 between H and $\tilde{B}$. By (I. 1.16) H_1 is a D-subgroup of E(A). But this leads to a contradiction since we would get a non-trivial linear Δ_0-subgroup $(H_1/H)_{\Delta 1}$ embedded into the abelian Δ_0-group $(E(A)/H)_{\Delta} = \Gamma$, cf. (3.10). We proved the equality $L(H)^{\perp} = S_{DR}(E(A)/H)_a = DH^0(\Omega^1_{A/\mathcal{U}})$ from which we get that $\dim \Gamma = \dim_{\mathcal{U}}(DH^0(\Omega^1_{A/\mathcal{U}}))$ and we also get the unicity of Γ. For the existence of a Zariski dense group Γ in $\hat{A}$ with no non-trivial linear representation we simply let $\Gamma = (E(A)/H)_{\Delta}$ where H is the maximum linear algebraic D-subgroup of E(A) and use (3.14).

(3.19) COROLLARY. Let A be an abelian $\mathcal{U}$-variety. Then the following are equivalent (notations as in (3.17), (3.18)):

1) $E(A)_{\Delta}$ is an abelian Δ_0-group.
2) $A^{\#} = E(A)_{\Delta}$.
3) $\dim A^{\#} = 2\dim A$
4) $DH^0(\Omega^1_{A/\mathcal{U}}) = H^1_{DR}(A)$.

(3.20) EXAMPLE. Assume $\dim A = 1$ (A an elliptic curve). Then only two possibilities may occur for $DH^0(\Omega^1_{A/\mathcal{U}})$ namely:

a) $DH^0(\Omega^1_{A/\mathcal{U}}) = H^0(\Omega^1_{A/\mathcal{U}})$.

b) $DH^0(\Omega^1_{A/\mathcal{U}}) = H^1_{DR}(A)$.

In case a) we get by (III. 3.4) that A is defined over $\mathcal{K}$, say $A \simeq A_0 \otimes \mathcal{U}$, A_0 an abelian $\mathcal{K}$-variety. Then clearly $A^{\#} = \check{A}_0 = A_0(\mathcal{K}) \subset A(\mathcal{U})$.

n case b) we get by (3.19) that $A^{\#} = E(A)_{\Delta}$ and A does not descend to $\mathcal{K}$.

Note that in the latter case $A^{\#}$ is the unique Δ_o-subgroup of dimension 2 of $\hat{A}$ (indeed, y (3.14) for any Δ_o-subgroup Γ of $\hat{A}$ of dimension 2, $G(\Gamma)$ is an extension of A by G_a; it cannot e a trivial extension by (IV. 3.1) so it must be the universal extension E(A)).

As an application of this unicity remark consider once again the Δ_o-subgroup Γ of $\hat{A}$ where A = Zariski closure in $\mathbf{P}^2$ of $Z(y^2 - x(x-1)(x-t)) \subset \mathcal{U}^2$, $t' = 1$) obtained by taking the Δ-closure in $\hat{\mathbf{P}}^2$ of

$$Z(y^2 - x(x-1)(x-t), -y^3 - 2(2t-1)(x-t)^2x'y + 2t(t-1)(x-t)^2(x''y - 2x'y'))$$

s in Example 3b) in (1.12). Since dim $\Gamma = 2$, by the unicity remark above we must have $\Gamma = A^{\#} = E(A)_{\Delta}$ so $G(\Gamma) = E(A)$ viewed as an algebraic D-group via the unique lifting of $\delta \in \mathrm{Der}\,\mathcal{U}$ to P(E(A)) whose explicit formula can be obtained as explained in the Example after (III. 2.19).

It would be nice to "see" the equality $\Gamma = A^{\#}$ directly, without using the above "unicity emark". In particular since $\mathcal{U}\langle\Gamma\rangle = \mathcal{U}(x,y,x')$ and $\mathcal{U}(E(A)) = \mathcal{U}(x,y,\xi)$ (with ξ as in (III. 2.19)) it vould be nice to "see" the formula expressing ξ as a rational function in x,y,x'.

(3.21) REMARK. More generally it would be interesting to dispose for any abelian $\mathcal{U}$-variety A defined explicitely by some algebraic equations in the projective space $\mathbf{P}^n$ of a method to get algebraic differential equations defining $A^{\#}$ inside A.

(3.22) REMARK. If we fix an abelian $\mathcal{U}$-variety A there may exist plenty of Zariski dense Δ_o-subgroups of $\hat{A}$; they may be of arbitrary dimension. For example take A an elliptic curve defined over $\mathcal{K}$ put $G = G_a^N \times A$ and assuming for simplicity that $\mathcal{U}$ in ordinary ($\Delta = \{\delta\}$) et δ^* be the trivial lifting of δ from $\mathcal{U}$ to G and $\theta \in H^o(T_A)$ be a non-zero global vector field. Then letting $\xi_1,\dots,\xi_N$ be a $\mathcal{U}$-basis of $X_a(G_a^N)$ and $\frac{\partial}{\partial\xi_1},\dots,\frac{\partial}{\partial\xi_N}$ be the corresponding basis of $L(G_a^N)$, put $\tilde{\delta} = \delta^* + \xi_1\theta + \xi_2\frac{\partial}{\partial\xi_1} + \dots + \xi_N\frac{\partial}{\partial\xi_{N-1}} \in \mathrm{Der}_{\mathcal{K}}\,\mathcal{U}(G)$. Then by (I. 1.9) $\tilde{\delta} \in P(G)$ so $\tilde{\delta}$ defines a structure of algebraic D-group on G. It is easy to check that there is no intermediate field between $\mathcal{U}(A)$ and $\mathcal{U}(G)$ preserved by $\tilde{\delta}$. Then G_{Δ} is an abelian Δ_o-group embedded in $\hat{A}$.

4. Moduli of Δ_o-groups

We propose ourselves to find descriptions of the set

$$\mathcal{M} = \{\Delta_o\text{-groups modulo isomorphisms}\}$$

or of remarkable subsets of it. We start with "set-theoretic descriptions". We shall continue by putting Δ-manifold structures on such sets.

By (2.8) the set $\mathcal{M}$ can be written as a disjoint union

$$\mathcal{M} = \coprod_G \mathcal{M}(G)$$

where G runs through the set of isomorphism classes of irreducible algebraic $\mathcal{U}$-groups and for any such G we denote $\mathfrak{M}(G)$ the set of isomorphism classes of Δ_0-groups Γ such that $G(\Gamma)^! \simeq G$.

So one reasonable thing to do (although as we shall see not the only one) is to fix G and try to describe $\mathfrak{M}(G)$.

Here are some easy consequences of our theory:

(4.1.) LEMMA

1) Assume G is affine. Then $\mathfrak{M}(G)$ is non-empty if and only if G is defined over $\mathcal{K}$.

2) Assume G is commutative. Then $\mathfrak{M}(G)$ is non-empty if and only if

$$\partial(\delta_i) \in \mathrm{Der}_{\mathcal{K}}^{S(G)}\mathcal{U} \qquad \text{for all } i.$$

3) Assume G has no non-trivial linear representation. Then $\mathfrak{M}(G)$ has at most one element.

Proof. Use (II. 3.31) and (III. 2.18).

Now we claim that for any irreducible algebraic $\mathcal{U}$-group G we have:

$$\mathfrak{M}(G) = P(G)^{int}/\mathrm{Aut}\,G$$

where $P(G)^{int}$ is the set of m-uples $(p_1,\dots,p_m)$ of pairwise commuting elements in P(G) such that $\partial(p_i) = \partial(\delta_i)$ for $1 \leq i \leq m$, while Aut G acts on $P(G)^{int}$ by the formula

$$(\sigma,(p_1,\dots,p_m)) \to (\sigma^{-1}p_1\sigma,\dots,\sigma^{-1}p_m\sigma), \quad \sigma \in \mathrm{Aut}\,G.$$

where σ stands also for the induced automorphism of $\mathcal{U}(G)$.
To check our claim note that by (2.8) $\mathfrak{M}(G)$ identifies with the set of Aut G - orbits in the set of algebraic D-group structures on G; but now we are done by (I. 1.1).

The above description of $\mathfrak{M}(G)$ is of course not satisfactory. To get a better picture assume for a moment that $\mathcal{K}$ is a field of definition for G so

$$G \simeq G_{\mathcal{K}} \otimes_{\mathcal{K}} \mathcal{U}$$

where $G_{\mathcal{K}}$ is some algebraic $\mathcal{K}$-group.
Let δ_i^* be the trivial lifting of $\partial(\delta_i)$ from $\mathcal{U}$ to G (as usual δ_i^* acts trivially on $\mathcal{K}(G_{\mathcal{K}})$). Then the map

$$P(G)^m \to P(G/\mathcal{U})^m$$

$$(p_1,\dots,p_m) \to (p_1 - \delta_1^*,\dots,p_m - \delta_m^*)$$

induces a bijection

$$\mathfrak{M}(G) \simeq P(G/\mathcal{U})^{int}/\mathrm{Aut}\,G$$

where

1) $P(G/\mathcal{U})^{int}$ is the set of m-uples $(\theta_1,\dots,\theta_m)$ of elements $\theta_i \in P(G/\mathcal{U})$ such that

$$\nabla_{\delta_i^*}\theta_j - \nabla_{\delta_j^*}\theta_i + [\theta_i,\theta_j] = 0 \qquad \text{for all } i,j$$

2) Aut G acts by the "Loewy-type" action [C1]:

$$(\sigma,(\theta_i)) \to (\sigma^{-1}\theta\sigma + \ell\nabla_{\delta_i^*}\sigma)_i$$

Here $(\nabla,\ell\nabla)$ is the logarithmic P(G)-connection on (Aut G, $P(G/\mathcal{U})$) defined in (I. 1.24) so recall that

$$\nabla_{\delta_i^*}\theta = [\delta_i^*,\theta],\ \ell\nabla_{\delta_i^*}\sigma = \sigma^{-1}\delta_i^*\sigma - \delta_i^* \qquad \text{for } \theta \in P(G/\mathcal{U}), \sigma \in \text{Aut } G.$$

This description is as unsatisfactory as the first one because everything is "hidden" by the "Δ-algebraic action" of Aut G; what one should be looking for is a description involving an algebraic (rather that a "Δ-algebraic") action. Such a description will be given in the Theorem below in a special case.

(4.2) THEOREM. Let $G_{\mathcal{K}}$ be an irreducible algebraic $\mathcal{K}$-group, $G = G_{\mathcal{K}} \otimes \mathcal{U}$ and assume P(G) contains a representative ideal V. Then we have an identification

$$\mathfrak{M}(G) = V^{int}/\text{Aut } G_{\mathcal{K}}$$

where

1) $V^{int} = \{(a_1,\dots,a_m) \in V^m;\ \nabla_{\delta_i^*}a_j - \nabla_{\delta_j^*}a_i + [a_i,a_j] = 0 \text{ for all } i,j\}$.

2) Aut $G_{\mathcal{K}}$ acts on V^{int} by the formula

$$(\sigma,(a_1,\dots,a_m)) \to (\sigma^{-1}a_1\sigma,\dots,\sigma^{-1}a_m\sigma)$$

Proof. The inclusion

$$V^{int} \to P(G/\mathcal{U})^{int}$$

induces a map

$$\phi : V^{int}/\text{Aut } G_{\mathcal{K}} \to P(G/\mathcal{U})^{int}/\text{Aut } G$$

We must check ϕ is a bijection. To check injectivity assume $\theta = (\theta_1,\dots,\theta_m)$ and $\theta' = (\theta'_1,\dots,\theta'_m)$ are Aut G-conjugate elements in $P(G/\mathcal{U})^{int}$, hence

$$\theta'_i = \sigma^{-1}\theta\sigma + \sigma^{-1}\delta_i^*\sigma - \delta_i^*$$

for some element $\sigma \in$ Aut G. Since V is an Aut G-invariant subspace of $P(G/\mathcal{U})$ we have

$$\theta'_i - \sigma^{-1}\theta\sigma \in V$$

Since P(G,fin) is an Aut G-invariant subspace of P(G) too, we have

$$\sigma^{-1}\delta_i^*\sigma - \delta_i^* \in P(G,fin)$$

Since $V \cap P(G,fin) = 0$ we get

$$a)\ \theta_i' = \sigma^{-1}\theta_i\sigma \qquad \text{for all } i$$

$$b)\ \delta_i^* = \sigma^{-1}\delta_i^*\sigma \qquad \text{for all } i.$$

From b) and (I. 3.6), 1) we get $\sigma \in \operatorname{Aut} G_{\mathcal{K}}$. By a) θ and θ' are $\operatorname{Aut} G_{\mathcal{K}}$-conjugate so ϕ is injective.

To check that ϕ is surjective take $\theta = (\theta_1,\dots,\theta_m) \in P(G/\mathcal{U})^{int}$. We may write

$$\theta_i = d_i + a_i$$

with $d_i \in P(G/\mathcal{U},fin)$, $a_i \in V$. We claim that $(d_1,\dots,d_m) \in P(G/\mathcal{U})^{int}$. Indeed we have in P(G) the equalities:

$$0 = [\delta_i^* + \theta_i, \delta_j^* + \theta_j] = [\delta_i^* + d_i, \delta_j^* + d_j] + [\delta_i^* + d_i, a_j] - [\delta_j^* + d_j, a_i] + [a_i, a_j]$$

Since V is an ideal in P(G) the last 3 terms of the right hand side of the above formula belong to V while the first term belongs to P(G,fin), because P(G,fin) is a Lie $\mathcal{U}/\mathcal{K}$-subalgebra of P(G), (IV. 1.2). We conclude as above that the first term vanishes, which proves our claim. By this claim we may consider the algebraic D-group structure on G defined by the map

$$P \longrightarrow P(G)$$

$$\delta_i \longmapsto \delta_i^* + d_i$$

cf. (I. 1.1). By (2.9) this algebraic D-group structure on G is split hence by (2.8) there exists $\sigma \in \operatorname{Aut} G$ such that

$$\sigma^{-1}(\delta_i^* + d_i)\sigma = \delta_i^* \qquad \text{for all } i.$$

We get that

$$\sigma^{-1}\theta_i\sigma + \sigma^{-1}\delta_i^*\sigma - \delta_i^* = \sigma^{-1}d_i\sigma + \sigma^{-1}a_i\sigma + \sigma^{-1}\delta_i^*\sigma - \delta_i^* = \sigma^{-1}a_i\sigma$$

hence θ is Aut G-conjugate to $(\sigma^{-1}a_1\sigma,\dots,\sigma^{-1}a_m\sigma) \in V^{int}$ and our Theorem is proved.

(4.3) The above theorem gives a rather satisfactory answer to the classification problem for linear Δ_0-groups. Indeed for any linear Δ_0-group Γ, $G = G(\Gamma)^!$ is affine by (2.18). By (II. 1.1) G is defined over $\mathcal{K}$ and by (II. 1.22) P(G) contains a representative ideal. So Theorem (4.2) applies to G! We get the following bijection

$$\mathfrak{M}^{lin} := \{\text{linear } \Delta_0\text{-groups modulo isomorphism}\} \simeq \coprod_G (V(G)^{int}/\operatorname{Aut} G_{\mathcal{K}})$$

where the disjoint union is taken over all isomorphism classes G of irreducible affine algebraic $\mathcal{U}$-groups defined over $\mathcal{K}$ and V(G) is a representative ideal in P(G).

Moreover note that if G is such that all its weights are killed by P(G) (e.g. if either the radical of G is nilpotent or the unipotent radical of G is commutative) we have by (II. 3.16) and (II. 3.22) that

$$V(G)^{int}/\mathrm{Aut}\,G_{\mathcal{K}} \simeq W_o(G)^{int}/\mathrm{Aut}\,G_{\mathcal{K}}$$

where $W_o(G)^{int} = \{(a_1,\dots,a_m) \in W_o(G);\ \delta_i^* a_j = \delta_j^* a_i$ for all i,j} (δ_i^* being here the trivial lifting of $\partial(\delta_i)$ from $\mathcal{U}$ to $W_o(G) = W_o(G_{\mathcal{K}}) \otimes \mathcal{U}$ and $\mathrm{Aut}\,G_{\mathcal{K}}$ acts on $W_o(G)^{int}$ via its action on $W_o(G_{\mathcal{K}})$.

(4.4) As already noted, by (2.10) a Δ_o-group Γ has no non-trivial linear representation if and only if $G(\Gamma^!)$ has the same property. Then (4.1) implies that we have a bijection

$$\mathcal{M}^{\#} := \left\{\begin{array}{l}\Delta_o\text{-groups with no non-trivial}\\ \text{linear representation modulo}\\ \text{isomorphism}\end{array}\right\} \simeq \left\{\begin{array}{l}\text{irreducible algebraic } \mathcal{U}\text{-groups G with}\\ \mathcal{O}(G) \simeq \mathcal{U} \text{ such that } \partial(\delta_i) \in \mathrm{Der}_{\mathcal{K}}^{S(G)}\mathcal{U}\\ \text{for all i, modulo isomorphism}\end{array}\right\}$$

(4.5) Assume $G = L \times A$ where L is an irreducible affine algebraic group and A is an abelian variety. Then by (IV. 3.1) $\mathcal{M}(G)$ is non-empty if and only if $\mathcal{K}$ is a field of definition for both L and A, say $A \simeq A_{\mathcal{K}} \otimes \mathcal{U}$, $L \simeq L_{\mathcal{K}} \otimes \mathcal{U}$. Assume this is the case and assume in addition for simplicity that P(L) = P(L,fin) (e.g. the radical of L is unipotent). Then we have by (IV. 3.1) and (4.2) that

$$\mathcal{M}(G) \simeq (X_a(L) \otimes L(A))^{int}/\mathrm{Aut}\,L_{\mathcal{K}} \times \mathrm{Aut}\,A_{\mathcal{K}}$$

where $(X_a(L)\otimes L(A))^{int} = \{(a_1,\dots,a_m) \in X_a(L)\otimes L(A);\ \delta_i^* a_j = \delta_j^* a_i\}$ (δ_i^* being the trivial lifting of $\partial(\delta_i)$ from $\mathcal{U}$ to

$$X_a(L)\otimes L(A) = (X_a(L_{\mathcal{K}})\otimes_{\mathcal{K}} L(A_{\mathcal{K}}))\otimes \mathcal{U}$$

and $\mathrm{Aut}\,L_{\mathcal{K}} \times \mathrm{Aut}\,A$ acts via its natural action on $X_a(L_{\mathcal{K}}) \otimes L(A_{\mathcal{K}})$.

(4.6) Let's give a description of

$$\mathcal{M}^{ab,\#} := \mathcal{M}^{ab} \cap \mathcal{M}^{\#} = \left\{\begin{array}{l}\text{abelian } \Delta_o\text{-groups with no non-trivial linear}\\ \text{representation modulo isomorphism}\end{array}\right\}$$

Indeed by (3.19) we have a decomposition into a disjoint union

$$\mathcal{M}^{ab,\#} = \coprod_{1\leq g\leq d\leq 2g} \mathcal{M}_{d,g}$$

where

$$\mathcal{M}_{d,g} = \left\{ \begin{array}{c} \Delta_0\text{-groups in } \mathcal{M}^{ab,\#} \text{ of dimension } d \text{ and} \\ \text{embedding dimension } g \end{array} \right\}$$

Moreover by (3.19) we have a bijection

$$\mathcal{M}_{d,g} \simeq \left\{ \begin{array}{l} \text{abelian } \mathcal{U}\text{-varieties } A \text{ of dimension } g \text{ with the property} \\ \qquad\qquad \dim_{\mathcal{U}} DH^0(\Omega^1_{A/\mathcal{U}}) = d \\ \text{modulo isomorphism} \end{array} \right\}$$

(4.7) Up to now we discussed only "set-theoretic" descriptions of "moduli sets". But there is an interesting possibility that some of these moduli sets (or some natural "coverings" or "pieces" of them) carry natural structures of Δ-manifolds.

For instance if G is as in Theorem (3.3) (with $\mathcal{U}$ ordinary, to simplify discussion) then $\mathcal{M}(G)$ is of the form

$$\mathcal{U}^n/A(\mathcal{K})$$

where A is an affine locally algebraic $\mathcal{K}$-group acting on $\mathcal{U}^n$ via a representation

$$A(\mathcal{K}) \xrightarrow{\rho} GL_n(\mathcal{K}) \subset GL_n(\mathcal{U})$$

where ρ is locally algebraic. Now A has infinitely many components in general so our task of finding Δ-manifold structures is hopeless unless we pass from $\mathcal{U}^n/A(\mathcal{K})$ to its "covering" $\mathcal{U}^n/A^0(\mathcal{K})$ (the fibres of the projection $\mathcal{U}^n/A^0(\mathcal{K}) \to \mathcal{U}^n/A(\mathcal{K})$ are then at most countable). The study of such quotients $\mathcal{U}^n/A^0(\mathcal{K})$ was part of a program initiated by B. Weisfeiler; some general theorems were proved by him and lots of conjectures can be made at least for A^0 reductive. In any case what one should expect is that, exactly as in the "non-differential" invariant theory only an "open subset" of $\mathcal{U}^n/A^0(\mathcal{K})$ can be equipped with a Δ-manifold structure.

Fortunately, for many linear algebraic groups G we dispose of a quite explicit description of the action of $\text{Aut}\, G_{\mathcal{K}}$ on a representative ideal (cf. (4.3)) this leading to a direct easy description of $\widetilde{\mathcal{M}}(G) = V/\text{Aut}^0 G_{\mathcal{K}}$; the latter turns out aposteriori to contain a "big" piece which is a Δ-manifold. We illustrate this with one example (the reader can consider more complicated examples and try to apply our theory to deal with them):

(4.8) THEOREM. We have a natural bijection

$$\mathcal{M}(G_a^N \times G_m^N) \simeq gl_N(\mathcal{U})/(GL_N(\mathbf{Z}) \times GL_N(\mathcal{K}))$$

Moreover the subset $\widetilde{\mathcal{M}}_0 = GL_N(\mathcal{U})/GL_N(\mathcal{K})$ of the set

$$\widetilde{\mathcal{M}}(G_a^N \times G_m^N) := gl_N(\mathcal{U})/GL_N(\mathcal{K})$$

dentifies (via logarithmic derivative $\ell\nabla_\delta : GL_N(\mathcal{U}) \to gl_N(\mathcal{U})$) with $gl_N(\mathcal{U})$. So $\tilde{\mathcal{M}}_0$ has a natural tructure of Δ-affine space $\mathcal{U}^{N^2}$.

Proof. In notations of (II. 3.1), (II. 3.12) we clearly have:

$$W_o(G) = W(G) = \mathrm{Hom}(\mathbf{Z}^N, \mathcal{U}^N) = gl_N(\mathcal{U})$$

vhile $\mathrm{Aut}(G_a^N \times G_m^N)_{\mathcal{K}} = GL_N(\mathbf{Z}) \times GL_N(\mathcal{K})$. By (4.3) we get our formula for $\mathcal{M}(G_a^N \times G_m^N)$. The est follows from Kolchin's surjectivity of $\ell\nabla_\delta$ (cf. (I. 2.3)).

(4.9) We close our search for "Δ-manifold structure on moduli sets" by considering belian Δ_o-groups with no non-trivial linear representation, i.e. the set $\mathcal{M}^{ab,\#}$, cf. (4.6).

Once again the example of abelian varieties from algebraic geometry teaches us that if one xpects moduli one has to introduce polarizations. In what follows by a polarized abelian Δ_o-group with level n structure we will understand the equivalence class of a triple

$$(\Gamma \to \hat{A}, \lambda, \eta)$$

where Γ is an abelian Δ_o-group, $\Gamma \to \hat{A}$ is a Zariski dense embedding (A an abelian $\mathcal{U}$-variety), λ s a polarization on A and η is a level n structure on A; the triples $(\alpha_i : \Gamma_i \to \hat{A}_i, \lambda_i, \eta_i)$, i = 1,2 are alled equivalent if $\Gamma_1 = \Gamma_2 = \Gamma$ and there is an isomorphism $\sigma : A_1 \to A_2$ such that $\alpha_1 = \alpha_2, \sigma^* \lambda_2 = \lambda_1$ and $\sigma(\eta_1) = \eta_2$. A triple as above will be said to be associated to Γ. There is n obvious notion of isomorphism of such objects. Moreover a triple as above will be called rincipally polarized if λ is a principal polarization (see [Mu] for backgraound). Note that given n abelian Δ_o-group Γ there are at most countably many triples as above associated to Γ, nodulo isomorphism, cf. (3.15).

Let $\mathcal{M}_{d,g,n}$ be the set of isomorphism classes of principally polarized abelian Δ_o-groups vith level n-structure having the following properties:

1) they have no non-trivial linear representation,

2) they have dimension d,

3) they have embedding dimension g.

In what follows we shall prove tht these $\mathcal{M}_{d,g,n}$ (for $g \geq 2$ and $n \geq 3$) have natural tructures of Δ-manifolds. They are non-empty only if $g \leq d \leq 2g$.

The case g = 1 was left aside just because it is too easy; we leave it to the reader, cf. 3.20). On the contrary the presence of level n-structures ($n \geq 2$) is a technical condition which e could not get rid of. Finally we expect that the case of non-principal polarizations can be orked with along the same lines as we do for principal ones; principal polarizations simplify iscussion at a few points.

We will prove in fact much more namely that $\mathcal{M}_{d,g,n}$ can be identified with certain ubsets of the corresponding moduli spaces of abelian varieties which are locally closed in the Δ – topology of these moduli spaces (of course they will be not locally closed in the usual Zariski topology). To state our main result let's make some notations. Denote by $\mathcal{A}_{g,n}$ be the moduli space of principally polarized abelian $\mathcal{U}$-varieties with level n structure [Mu] (viewed as an $\mathcal{U}$-variety). Note that $\mathcal{A}_{g,n}$ is defined over $\mathcal{K}$ [Mu] so we have $\mathcal{A}_{g,n} = \mathcal{A}^0_{g,n} \otimes \mathcal{U}$, $\mathcal{A}^0_{g,n}$ a $\mathcal{K}$-variety. Then for $g \leq d \leq 2g$ the natural maps

$$\mathcal{M}_{d,g,n} \to \mathcal{A}_{g,n}$$
$$(\Gamma \to \hat{A},\lambda,\eta) \to (A,\lambda,\eta)$$

are injective by (3.18) and their images $\mathcal{A}_{d,g,n}$ are disjoint and described as

$$\mathcal{A}_{d,g,n} = \left\{ \begin{array}{l} \text{points in } \mathcal{A}_{g,n} \text{ which correspond to abelian} \\ \text{varieties A such that } \dim_{\mathcal{U}} DH^0(\Omega^1_{A/\mathcal{U}}) = d \end{array} \right\}$$

cf. (3.18), (4.6); note that $\mathcal{M}_{d,g,n}$ and $\mathcal{A}_{d,g,n}$ can be identified as sets. Here is our main result.

(4.10) THEOREM. In notations above, if $g \geq 2$ and $n \geq 3$ then:

1) For any $g \leq d \leq 2g$ the sets $\mathcal{A}_{d,g,n}$ are non-empty and the sets

$$\mathcal{A}_{g,g,n} \cup \mathcal{A}_{g+1,g,n} \cup \dots \cup \mathcal{A}_{d,g,n}$$

are Δ-closed in $\mathcal{A}_{g,n}$.

2) We have $\mathcal{A}_{g,g,n} = \mathcal{A}^0_{g,n}$ and $\mathcal{A}_{g,g,n} \cup \dots \cup \mathcal{A}_{2g,g,n} = \mathcal{A}_{g,n}$.

(4.11) Note that the equality $\mathcal{A}_{g,g,n} = \mathcal{A}^0_{g,n}$ follows from (III. 3.4). The rest of the assertions require a preparation.

(4.12) Let Y be any $\mathcal{U}$-variety; denote by $TY = V(T_Y) = \mathrm{Spec}(S(\Omega^1_{Y/\mathcal{U}}))$. It coincides with the tangent bundle of Y (viewed as a variety) in case Y is smooth. Assume for simplicity this is the case. Assume moreover that Y is defined over $\mathcal{K}$ and fix an isomorphism $Y \simeq Y_0 \otimes \mathcal{U}$. We claim that for each $p \in P$ there is a natural map of Δ-manifolds

$$\nabla_p : \hat{Y} \to TY$$

To define it, it is sufficient to define a D-map

$$\nabla_p : Y^\infty \to (TY)^\infty$$

hence by (0.9) and (I. 2.12) it is sufficient to define for any (reduced) D-algebra R maps $\nabla_p : Y(R) \to (TY)(R)$ behaving functorially in R. We define them as follows. For any morphism $f : \mathrm{Spec}\, R \to Y$ still denote by f the map induced between the corresponding sheaves of algebras, denote by $p_R \in \mathrm{Der}_{\mathcal{K}}(R)$ the multiplication by $p \in P$ and by $p^* \in \mathrm{Der}_{\mathcal{K}}(\mathcal{O}_Y)$ the trivial lifting of $\partial(p)$ from $\mathcal{U}$ to $Y = Y_0 \otimes \mathcal{U}$; then the difference

$$p_R f - fp^*$$

ill be an $\mathcal{U}$-derivation from $\mathcal{O}_Y$ to $f_*\mathcal{O}_{\mathrm{Spec}\,R}$ hence will induce an $\mathcal{O}_Y$-linear map $: \Omega^1_{Y/\mathcal{U}} \to f_*\mathcal{O}_{\mathrm{Spec}\,R}$. The pair (f,θ) then defines an $\mathcal{U}$-algebra map $S(\Omega^1_{Y/\mathcal{U}}) \to f_*\mathcal{O}_{\mathrm{Spec}\,R}$ ence a morphism $\mathrm{Spec}\,R \to TY$ which we call $\nabla_p f$.

ore generally for any family $p_1, p_2, \ldots, \in P$ we have a morphism of Δ-manifolds $(\nabla_{p_1}, \nabla_{p_2}, \ldots) : \hat{Y} \to (TY \times_Y TY \times \ldots)\hat{}$.

(4.13) Clearly, to prove (4.10) we need a method to study the "variation" of the D-module tructure of $H^1_{DR}(A)$ as A varies in a moduli space. Remarkably this can be done by translating he D-module properties of $H^1_{DR}(A)$ in terms of the Gauss-Manin connection of any deformation of A which is defined over $\mathcal{K}$". This is what we do next.

(4.14) Let $\Pi_0 : X_0 \to Y_0$ be a smooth projective morphism of smooth $\mathcal{K}$-varieties, assume s fibres are connected and let $\Pi : X \to Y$ be obtained from $X_0 \to Y_0$ by tensorization with . et moreover $y \in Y(\mathcal{U})$ be any $\mathcal{U}$-point of Y and X_y be the fibre of $X \to Y$ at y.

By (III. 1.1) we dispose of a Gauss-Manin connection which is an $\mathcal{U}$-linear map.

$$\text{(4.14.1)} \qquad \nabla^{X_y} : \mathrm{Der}_{\mathcal{K}}\,\mathcal{U} \longrightarrow \mathrm{Hom}_{\mathcal{K}}(H^1_{DR}(X_y), H^1_{DR}(X_y)), \quad d \mapsto \nabla^{X_y}_d$$

On the other hand one can consider the relative de Rham sheaf $H^1_{DR}(X/Y) := R^1\Pi_*(\Omega^*_{X/Y})$; recall from [Ka] that this is a locally free $\mathcal{O}_Y$-module and its formation is ompatible with any base change, in particular $H^1_{DR}(X/Y) \otimes \mathcal{U}(y) \simeq H^1_{DR}(X_y)$($\mathcal{U}(y)$ = residue field f Y at y). Denote by $H^1_{DR}(X/Y)_y$ the stalk at y and by

$$r_y : H^1_{DR}(X/Y)_y \to H^1_{DR}(X_y)$$

he reduction modulo the maximal ideal m_y of $\mathcal{O}_{Y,y}$. By [Ka] we also dispose of a Gauss-Manin onnection

$$\text{(4.14.2)} \qquad \nabla^{X/Y} : T_Y = \check{\Omega}^1_{Y/\mathcal{U}} \to \underline{\mathrm{Hom}}_{\mathcal{U}}(H^1_{DR}(X/Y), H^1_{DR}(X/Y))$$

defined in the same way as the one in (III. 1.1)) inducing at y an $\mathcal{O}_{Y,y}$ - linear map

$$\text{(4.14.3)} \qquad \nabla^{X/Y,y} : \mathrm{Der}_{\mathcal{U}}(\mathcal{O}_{Y,y}) \to \mathrm{Hom}_{\mathcal{U}}(H^1_{DR}(X/Y)_y, H^1_{DR}(X/Y)_y), \; \theta \mapsto \nabla^{X/Y,y}_\theta$$

he map (4.14.1) should be thought of as an "internal" Gauss-Manin connection while (4.14.2) as n "external" one.

Note that we can see T_yY as a subset of TY (cf. (4.12)); it is of course the fibre of the rojection $TY \to Y$ at y. In particular denoting by $f_y : \mathrm{Spec}\,\mathcal{U} \to Y$ the morphism defining $\in Y(\mathcal{U})$ and also the induced morphism $f_y : \mathcal{O}_{Y,y} \to \mathcal{U}$ we may consider by (4.12) for any $p \in P$ he $\mathcal{U}$-point $\nabla_p f_y \in T_yY \subset TY$ defined by

$$\text{(4.14.4)} \qquad \nabla_p f_y = \partial(p) f_y - f_y p^* \in \mathrm{Der}_{\mathcal{U}}(\mathcal{O}_{Y,y}, \mathcal{U}) = T_yY$$

where p^* is the trivial lifting of $\partial(p)$ from $\mathcal{U}$ to Y, viewed here as an element of $\mathrm{Der}_{\mathcal{K}}(\mathcal{O}_{Y,y})$. Choose moreover any derivation $\theta \in \mathrm{Der}_{\mathcal{U}}(\mathcal{O}_{Y,y})$ "lifting" $\nabla_p f_y$ i.e. with the property that

(4.14.5) $$\nabla_p f_y = f_y \theta$$

Finally, let p^{**} be the trivial lifting of $\partial(p)$ from $\mathcal{U}$ to X. It induces an obvious $\mathcal{K}$-linear endomorphism ∇_{p^*} of $H^1_{DR}(X/Y)$ (any derivation of $\mathcal{O}_X$ lifting a derivation of $\mathcal{O}_Y$ does!).

The following lemma shows that the "internal" data can be recaptured from the "external" ones. Philosophically this is possible because of our hypothesis that the "deformation" X/Y of X_y is defined over $\mathcal{K}$, hence is "sufficiently rich".

(4.15) LEMMA. The following diagram is commutative

$$\begin{array}{ccc} H^1_{DR}(X/Y)_y & \xrightarrow{\nabla_{p^*} + \nabla_\theta^{X/Y,y}} & H^1_{DR}(X/Y)_y \\ \Big\downarrow r_y & & \Big\downarrow r_y \\ H^1_{DR}(X_y) & \xrightarrow{\nabla^{X_y}_{\partial(p)}} & H^1_{DR}(X_y). \end{array}$$

Proof. Put $\tilde{Y} = \mathrm{Spec}\,\mathcal{O}_{Y,y}$, $\tilde{X} = X \times_Y \tilde{Y}$, cover $\tilde{X}$ by open affine subsets $\tilde{X}_i = \mathrm{Spec}\,A_i$, lift θ to derivations $\theta_i \in \mathrm{Der}_{\mathcal{U}} A_i$, consider the cocycle $(\theta_j - \theta_i) \in C^1((\tilde{X}_i)_i, T_{\tilde{X}/\tilde{Y}})$ and reduce it modulo m_y to get a cocycle

$$\overline{(\theta_j - \theta_i)} \in C^1(X_{y,i}, T_{X_y})$$

where $X_{y,i} = X_y \cap \tilde{X}_i$ and the upper bar means "reduction modulo m_y" (note that θ_j cannot be reduced modulo m_y, so one cannot speak about $\overline{\theta}_j$, because θ_j may not preserve m_y; only the differences $\theta_j - \theta_i$ preserve it, indeed they vanish on it). By (4.14.4) and (4.14.5) we get the equality

$$\partial(p) f_y = f_y(p^* + \theta)$$

which shows in particular that $p^* + \theta$ preserves m_y. Hence its liftings $p^{**} + \theta_i \in \mathrm{Der}_{\mathcal{K}} A_i$ preserve m_y so they can be reduced modulo m_y to get derivations

$$\overline{p^{**} + \theta_i} \in \mathrm{Der}_{\mathcal{K}} \overline{A}_i$$

where $\overline{A}_i = \mathcal{O}(X_{y,i})$. We get the equality

$$\overline{\theta_j - \theta_i} = \overline{(p^{**} + \theta_j)} - \overline{(p^{**} + \theta_i)}$$

Now choose a representative (ω_i, x_{ij}) of a class $\eta \in H^1_{DR}(X/Y)_y = H^1_{DR}(\tilde{X}/\tilde{Y})$ where ω_i are 1-forms on $\tilde{X}_i$ and $x_{ij} \in \mathcal{O}(\tilde{X}_i \cap \tilde{X}_j)$. Then $r_y(\nabla_\theta^{X/Y,y})\eta$ is represented by

$$(*)\qquad (\overline{\mathrm{Lie}_{\theta_i}\,\omega_i}, \overline{(\theta_j - \theta_i)}\wedge\overline{\omega}_j + \overline{\theta_i x_{ij}})$$

Moreover $(\nabla^{X_y}_{\partial(p)})r_y\eta$ is represented by

$$(**)\qquad (\mathrm{Lie}_{\overline{p^{**}+\theta_i}}\,\overline{\omega}_i, (\overline{(p^{**}+\theta_j)} - \overline{(p^{**}+\theta_i)})\wedge\omega_j + \overline{(p^{**}+\theta_i)}(\overline{x}_{ij}))$$

while $r_y\nabla_{p^*}\eta$ is represented by

$$(***)\qquad (\overline{\mathrm{Lie}_{p^{**}}\,\omega_i}, \overline{p^{**}x_{ij}})$$

and our lemma follows.

Now for any p_i in a commuting basis $p_1,\dots,p_m$ of P let's fix a θ_i as in (4.14) i.e. with $p_i f_y = f_y\theta_i$.

(4.16) LEMMA. In notations of (4.14) and (3.17) let $\tilde{D}$ be the subring of $\mathrm{End}_{\mathcal{K}}(H^1_{DR}(X/Y)_y)$ generated by $\mathcal{U}$ and the maps $\nabla_i := \nabla_{p_i^*} + \nabla^{X/Y,y}_{\theta_i}$, $1 \le i \le m$. Then we have the equality

$$DH^0(\Omega^1_{X_y}/\mathcal{U}) = r_y(\tilde{D}\,H^0(\Omega^1_{X/Y})_y)$$

Proof. Let D_N be the $\mathcal{U}$-linear subspace of D generated by $\mathcal{U}$ and the products $\partial_{i_1}\partial_{i_2}\dots\partial_{i_k}$ $(k \le N)$ and let $\tilde{D}_N$ be the $\mathcal{U}$-linear subspace of $\tilde{D}$ generated by $\mathcal{U}$ and the products of the form

$$\nabla_{i_1}\nabla_{i_2}\dots\nabla_{i_k} \qquad (k \le N)$$

We check by induction on N that

$$(*)_N \qquad D_N H^0(\Omega^1_{X_y}/\mathcal{U}) = r_y(\tilde{D}_N H^0(\Omega^1_{X/Y})_y)$$

For $N = 0$ this is just the base change theorem. Assume equality holds for $N-1$ and let $\eta \in H^0(\Omega^1_{X/Y})_y$, $\tilde{d} \in \tilde{D}_{N-1}$ and i any index between 1 and m. Moreover let's agree to denote $\nabla^{X_y}_{\partial(p_i)}$ again by ∇_i (no confusion will arrise with the previous notations!). Then we have by (4.15):

$$r_y(\nabla_i\tilde{d}\,\eta) = (\nabla_i)(r_y\tilde{d}\,\eta)$$

But by induction hypothesis

$$(\nabla_i)(r_y\tilde{d}\,\eta) \in \nabla_i D_{N-1}H^0(\Omega^1_{X_y}/\mathcal{U}) \subset D_N H^0(\Omega^1_{X_y}/\mathcal{U})$$

So we get

$$r_y(\nabla_i\tilde{d}\,\eta) \in D_N H^0(\Omega^1_{X_y}/\mathcal{U})$$

which proves the inclusion "$\supset$" in $(*)_N$. The inclusion "$\subset$" follows by an entirely similar argument.

(4.17) To formulate the next lemma let's introduce some new constructions. In notations from (4.14) let Y' be an open subset of Y whose tangent bundle is trivial and fix a trivialisation $T_{Y'} \simeq \mathcal{O}^N_{Y'}$. Associated to this trivialisation consider the natural map

$$\tau : TY' \to H^0(Y',T_Y)$$

(for each vector v in some T_yY consider the vector field on Y' obtained by displacing v in the other points of Y' via the trivialisation of $T_{Y'}$). Then for any point

$$(y,v) = (y,v_1,\dots,v_m) \in \underbrace{TY \underset{Y}{\times} \dots \underset{Y}{\times} TY}_{m \text{ factors}},\ y \in Y',\quad v_i \in T_yY$$

we can consider the $\mathcal{K}$-endormorphism $\nabla_1(y,v),\dots,\nabla_m(y,v)$ of $H^1_{DR}(X/Y)_y$ defined by the formula

$$\nabla_i(y,v) = \nabla_{p_i^*} + \nabla^{X/Y,y}_{\tau(y,v_i)} \qquad 1 \le i \le m$$

and define $\tilde{D}_{y,v}$ to be the subring of $\mathrm{End}_{\mathcal{K}}(H^1_{DR}(X/Y)_y)$ generated by $\mathcal{U}$ and $\nabla_i(y,v)$, $1 \le i \le m$. Then we have

(4.18) LEMMA. In notations from (4.17) and (4.14) for any integer d the set

$$\mathrm{Dis}_d = \{(y,v) \in TY \underset{Y}{\times} \dots \underset{Y}{\times} TY;\ y \in Y',\ \dim_{\mathcal{U}} r_y(\tilde{D}_{y,v} H^0(\Omega^1_{X/Y})_y) \le d\}$$

is Δ-closed in $TY \underset{Y}{\times} \dots \underset{Y}{\times} TY$.

Proof. We may assume Y = Y'; shrinking Y (the question is local) we may assume $H^0(\Omega^1_{X/Y})$ is free and a direct summand in $H^1_{DR}(X/Y)$ which is also free. We also may write $TY = Y \times \mathcal{U}^N$, and choose a frame of vector fields $w_1,\dots,w_N$ on Y. The map $\tau : TY' = Y' \times \mathcal{U}^N \to H^0(Y',T_Y)$ has then the form

$$\tau(y,\lambda_1,\dots,\lambda_N) = \Sigma\lambda_i w_i$$

The elements of $TY \underset{Y}{\times} \dots \underset{Y}{\times} TY$ are then pairs (y,λ) where $\lambda = (\lambda_{ij})$ is an $m \times N$ matrix with elemetns in $\mathcal{U}$ and

$$\nabla_i(y,\lambda) = \nabla_{p_i^*} + \sum_{j=1}^{N} \lambda_{ij}\nabla_{w_j}$$

Finally, fix a basis $e_1,\dots,e_n$ of $H^0(\Omega^1_{X/Y})$ as an $\mathcal{O}_Y$-module which extends to a basis $e_1,\dots,e_n,\dots,e_k$ of $H^1_{DR}(X/Y)$; with respect to this basis we may write

$$\nabla_{p_i^*} = p_i^* + M_i$$

$$\nabla_{w_i} = w_j + N_j$$

here M_i, N_j are matrices with entries in $\mathcal{O}(Y)$.
ll we have to check is that for any fixed matrix $(a_{i\alpha})$ of elements in $\mathcal{O}(Y)$ and any integer r the ollowing set is Δ-closed:

$$(*)\qquad \{(y,\lambda) \in Y \times \mathcal{U}^{mN};\ \dim_{\mathcal{U}} \langle r_y d_\alpha(\Sigma a_{i\alpha} e_i);\ |\alpha| \leq r\rangle \leq d\}$$

here $\langle\ \rangle$ means "$\mathcal{U}$-linear span", $\alpha = (\alpha_1, \alpha_2, \dots)$ are sequences of integers of length $|\alpha|$ and

$$d_\alpha := \nabla_{\alpha_1} \nabla_{\alpha_2} \dots = (p^*_{\alpha_1} + M_{\alpha_1} + \sum_j \sigma_{\alpha_1 j}(w_j + N_j))(p^*_{\alpha_2} + M_{\alpha_2} + \sum_j \lambda_{\alpha_2 j}(w_j + N_j)) \dots$$

low the coefficients of the e_i's in $d_\alpha(\Sigma a_{i\alpha} e_i)$ are polynomials in $\partial(p_{i_1}) \dots \partial(p_{i_k})\lambda$ whose oeffcients are (fixed) functions in $\mathcal{O}(Y)$. The conditions defining $(*)$ are obtained by the anishing of certain determinants whose entries are the coefficients of the e_i's in $d_\alpha(\Sigma a_{i\alpha} e_i)$ valuated at various points $y \in Y'$. This closes the proof.

(4.19) Let's check assertion 1) in Theorem (4.10). As well known the moduli space $= \mathcal{A}_{g,n}$ is smooth ($n \geq 3$!) and there is an abelian scheme $\Pi: X \to Y$ (the "universal family"). loreover X, Y, Π are defined over the prime field, in particular over $\mathcal{K}$ so we are in the situation f (4.14). The statement of (4.10), 1) is local on Y so we may replace Y by some affine subset here the tangent bundle is trivial. Noting that in notations of (4.17), $\tau(y, v_i) \in H^0(T_y)$ lifts (by ts very definition) the tangent vector $v_i \in T_y Y$ we get by (4.16) that

$$DH^0(\Omega^1_{X_y/\mathcal{U}}) = r_y(\tilde{D}(y, \nabla_{p_1} f_y, \dots, \nabla_{p_m} f_y)^{H^0(\Omega^1_{X/Y})_y})$$

t follows by (4.9) that $\mathcal{A}_{d,g,n}$ is the inverse image of Dis_d from (4.18) via the map:

$$Y(\mathcal{U}) \to (TY \underset{Y}{\times} \dots \underset{Y}{\times} TY)(\mathcal{U})$$

$$y \mapsto (\nabla_{p_1} f_y, \dots, \nabla_{p_m} f_y)$$

he latter map being a map of Δ-manifolds (4.12), closedness of $\mathcal{A}_{d,g,n}$ in the Δ-opology follows from (4.18).

To conclude the proof of (4.10) it remains to produce for each d with $g \leq d \leq 2g$ an belian variety A_d belonging to $\mathcal{A}_{d,g,n}$. For this let E be an elliptic curve not defined over the ield of constants of δ_1, say, let F be an elliptic curve defined over $\mathcal{K}$ and put $_d = E^{d-g} \times F^{2g-d}$.

(4.20) It would be interesting to know whether Theorem (4.10) holds for $1 \leq n \leq 2$ especially for $n = 1$ since "level 1-structure" means simple "no level structure"!). So assume ≤ 2. Then assertion 3) in (4.10) still holds for identical reasons. Moreover the sets

$$(*)\qquad \mathcal{A}_{g,g,n} \cup \mathcal{A}_{g+1,g,n} \cup \dots \cup \mathcal{A}_{d,g,n}$$

re the projections via $\mathcal{A}_{g,3} \to \mathcal{A}_{g,n}$ of the corresponding sets for $n = 3$. The latter are known to be Δ-closed by (4.10). But in spite of the fact that the map $\mathcal{A}_{g,3} \to \mathcal{A}_{g,n}$ is finite it is not clear to us whether it takes Δ-closed sets into Δ-closed sets. Note that a "Chevalley constructibility theorem" holds in our context [B_2][B_3] so the sets (*) are at least constructible in the Δ-topology!

(4.21) The loci $\mathcal{A}_{d,g,n}$ deserve further study. They stratify (in the Δ-topology) the moduli space in a non-trivial way and bare a formal resemblance with the degeneracy loci in "Brill-Noether theory". It would be interesting to compute their Δ-dimension polynomials [K_1], to investigate their irreducibility or specialisation properties (e.g. is $\mathcal{A}_{d,g,n}$ contained in the Δ-closure of $\mathcal{A}_{d+1,g,n}$?).

APPENDIX A. LINK WITH MOVABLE SINGULARITIES

In this appendix we discuss the relation between algebric D-groups and Δ-function fields with no movable singularity (NMS) in the sense of [B_1] (recall that by the latter we mean an extension of D-fields $F \subset E$, where $D = F[P]$, P a Lie F/k-algebra, such that $E = F\langle V\rangle$ where V is some projective D-variety. For the understanding of the discussion below, a certain familiarity with [B_1] is preferable. We shall assume from now on that all Lie K/k-algebras occuring have a finite commuting basis and all fields have characteristic zero; K will be always algebraically closed. There are easy examples showing that if P is a Lie K/k-algebra, $D = K[P]$ and G is an algebraic D-group then the extension $K \subset K\langle G\rangle$ need not have (NMS); for instance P. Cassidy proved that this is the case with $G = \mathrm{Spec}(K\{y\}_y/[yy'' - (y')^2])$ (note that $G^! = G_a \times G_m$). On the other hand we can prove:

(A.1) PROPOSITION. Let G be an irreducible algebraic D-group all of whose weights are P-constant. Then there exist a Picard-Vessiot extension K_1/K and an intermediate D-field $K_1 \subset E \subset K_1\langle G\rangle$ such that E/K_1 is split (i.e. $E = K_1(E^P)$) and $K_1\langle G\rangle/E$ is regular (i.e. E is algebraically closed in $K_1\langle G\rangle$) and generated by exponential elements (i.e. elements of $K_1\langle G\rangle$ whose logarithmic derivatives belong to E).

For the concept of Picard-Vessiot extension see [K_1], Chapter 5.

We will also prove the following proposition (where we recall that an element x of a D-field extension E of F is called a Picard-Vessiot element if x is contained in a finite dimensional D-submodule of E).

(A.2) PROPOSITION. Any D-field extension E/F (with F algebraically closed in E, $D = F[P]$, P a Lie F/k-algebra) is generated by Picard-Vessiot elements has (NMS).

So we get by [B_1] p. 103:

(A.3) COROLLARY. In notations of (A.1) all three extensions

$$K \subset K_1 \subset E \subset K_1\langle G\rangle$$

have (NMS).

(A.4) Proof of (A.1). Denote by $\delta_1,\dots,\delta_m$ a commuting basis of P and let p_i be the image of δ_i in P(G). Since we assumed all weights of G are P-constant we get (in notations of (II. 3.16)) that $a_i := \ell\nabla_{\delta_i} \in W_o(G)$ so by loc.cit. we may write

$$p_i = \delta_i^* + d_i + E_{a_i}$$

where $\delta_i^* \in P(G)$ is the trivial lifting of $\partial(\delta_i)$ from K to $G = G_o \otimes_{K_o} K$ ($K_o = K^D$, G_o a K_o-group), $d_i \in P(G/K,\text{fin})$ and $a \mapsto E_a$ is the map from $W_o(G)$ to P(G) appearing in (II. 3.16). Exactly as in the proof of (V. 4.2) the derivations $(\delta_i^* + d_i)_i$ are parirwise commuting so by (I. .1) they define an algebraic D-group structure on G. By the proof of (I. 3.11) we can find an automorphism $\sigma \in \text{Aut}(G \otimes K_1)$ where K_1/K is a Picard-Vessiot extension such that

$$\sigma^{-1}(\delta_i^* + d_i)\sigma = \delta_i^*$$

so we may assume the images of δ_i in $P(G \otimes K_1)$ have the form

$$\delta_i^* + E_{\sigma a_i}$$

Let U and H be as in (II. 3.21) and put $E = K_1(U)$; clearly E is preserved by $D_1 = K_1 \otimes_K D$ and E/K_1 is split. Moreover the extension $K_1\langle G\rangle/E$ is generated by elements of the form $1 \otimes y$ with

$$y \in \bigcup_{\chi} \mathcal{O}(H_o)^{\chi}$$

where $H = H_o \otimes_{K_o} K$, H_o a K_o-group, notations as in (II. 3.21) again) and any such $1 \otimes y$ is an exponential element in the extension $K_1\langle G\rangle/E$ because its logarithmic derivative with respect to p_i is $a_i(\chi) \otimes 1 \in E$ (if $y \in \mathcal{O}(H_o)^{\chi}$). Our proposition is proved.

(A.5) Proof of (A.2). Let E/F be a regular extension of D-fields generated by Picard-Vessiot elements. Then clearly a finite set of them suffices so $E = F(X)$, where $X = \text{Spec}\,A$ and A is some D-subalgebra of E, finitely generated as an F-algebra and locally finite as a D-module. Let $V \subset A$ be a finite dimensional D-submodule generating A as an F-algebra. Then the symmetric algebra SV has a natural structure of D-algebra and the closed embedding $X \to Y = \text{Spec}\,SV$ is a D-map. Give the polynomial algebra (SV)[t] a structure of D-algebra extension of SV by letting $pt = 0$ for all $p \in P$. Since D preserves the gradation on (SV)[t] it follows that $Q := \text{Proj}((SV)[t])$ becomes a D-variety and the open embedding $Y \to Q$ is a D-map. Let X be the Zariski closure of X in Q. By (0.14), 2) X is a D-subvariety of Q and this closes the proof.

Using (IV. 3.1) one can prove (in a way similar to (A.1) above):

(A.6) PROPOSITION. Let G be an algebraic D-group and assume its underlying K-group G is the product of a unipotent linear group by an abelian variety. Then there exist a Picard-Vessiot extension K_1/K (which by $[B_1]$ has (NMS)) and an intermediate D-field

$K_1 \subset E \subset K_1\langle G\rangle$ such that E/K_1 is regular and split (hence by $[B_1]$ has (NMS)) and $K_1\langle G\rangle/E$ has (NMS).

We close this Appendix by briefly inspecting the relation with Painlevé's property (cf. [GS], [Ha]). We need an analytic preparation:

(A.7) Let $f : X \to Y$ be a map of analytic complex manifolds and suppose we are given analytic commuting vector fields $\delta_1,\dots,\delta_m$ on Y lifting to some commuting vector fields $\hat\delta_1,\dots,\hat\delta_m$ on X. We say that f has the Painlevé property (with respect to these vector fields) if, upon letting $\tilde Y \to Y$ be the universal covering of Y, there is an analytic $\tilde Y$-isomorphism

$$X \times_Y \tilde Y \simeq Z \times \tilde Y$$

$Z = f^{-1}(y_0)$, $y_0 \in Y$, sending $\tilde\delta_i$ into δ^*_i where $\tilde\delta_i$ is the unique lifting of $\hat\delta_i$ from X to $X \times_Y \tilde Y$ and δ^*_i is the trivial lifting of δ_i from $\tilde Y$ to $Z \times \tilde Y$.

Note that if $\dim Y = m$ and if we throw away from Y the locus where $\delta_1,\dots,\delta_m$ do not generate the tangent space and from X the preimage of this locus (we assume that we are left with some non-empty open submanifolds!) then we get a foliation on X transverse to f and being a "feuilletage de Painlevé de 1re espece" in the sense of [GS], i.e. having the property that any path on Y starting from $y_0 \in Y$ can be lifted in the leaf passing through any point of $f^{-1}(y_0)$.

(A.8) Let's assume in what follows that K is an algebraically closed field containing $\mathbf{C}$ with tr.deg. $K/\mathbf{C} = m < \infty$ and put $P = \mathrm{Der}_{\mathbf{C}}K$, $D = K[P]$. Let $p_1,\dots,p_m$ be a commuting basis of P. Let moreover V be a D-variety. We say that V is a Painlevé D-variety if there is a cartesian diagram of $\mathbf{C}$ - schemes:

$$\begin{array}{ccc} U & \longrightarrow & V \\ {\scriptstyle f}\downarrow & & \downarrow \\ S & \xrightarrow[j]{} & \mathrm{Spec}\,K \end{array}$$

with U and S smooth $\mathbf{C}$ - varieties and j dominant and if there are commuting vector fields $\delta_1,\dots,\delta_m$ on S lifting to commuting vector fields $\hat\delta_1,\dots,\hat\delta_m$ on U such that

1) Each $\partial(p_i) \in \mathrm{Der}_{\mathbf{C}} K$ lifts δ_i and the image of each p_i in $\mathrm{Der}_{\mathbf{C}}\mathcal{O}_V$ lifts $\hat\delta_i$.

2) $f^{an} : U^{an} \to S^{an}$ has the Painlevé property (with respect to $\delta_1,\dots,\delta_m$ and $\hat\delta_1,\dots,\hat\delta_m$ viewed as analytic vector fields on S^{an}, U^{an}).

Finally let's say that a D-field extension E/K is a Painlevé extension if $E = K\langle V\rangle$ where V is some Painlevé D-variety.

The following hold:

a) Any smooth projective D-variety is a Painlevé D-variety; in particular, any extension E/K with (NMS) is a Painlevé extension. This is a trivial consequence of Ehresmann's theorem [J].

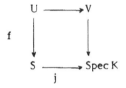

b) A consequence of the above remark and of our result in $[B_1]$ p. 103 is that any strongly normal extension of K is a Painlevé extension (as it has (NMS)).

c) Let G be an irreducible algebraic D-group. Then K<G>/K is a Painlevé extension (more precisely G is a Painlevé D-variety!). This follows from Hamm's result (II. 1.3), cf. [Ha].

d) Painlevé extensions need not have (NMS), for they may have "movable singularities hidden at infinity". For instance K. Okamoto's work (in Sem. F. Norguet, Février 1977) shows that the famous Painlevé second order equations "lead" to Painlevé extensions; these are not necessarily NMS extensions (as proved by Nishioka [N]).

APPENDIX B. ANALOGUE OF A DIOPHANTINE CONJECTURE OF S. LANG

The following result was proved in $[B_6]$ and illustrates the "diophantine flavour" of Δ_0-groups:

(B.1) THEOREM. Let A be an abelian $\mathcal{U}$-variety with $\mathcal{U}/\mathcal{K}$-trace zero, $X \subset A$ a subvariety and $\Gamma \subset \hat{A}$ a Δ_0-subgroup. Then there exist abelian subvarieties $B_1,\dots,B_m$ of A and points $a_1,\dots, a_m \in A$ such that

$$\Gamma \cap X \subset \bigcup_{i=1}^{m} (B_i + a_i) \subset X$$

The proof relies on analytic arguments (namely the Painlevé property in (A.8), c) plus a "Big Picard Theorem" due to Kobayashi-Ochiai). Using (B.1) we deduced in $[B_6]$ the following

(B.2) COROLLARY. Let A be an abelian variety over an algebraically closed field F of characteristic zero, assume A has F/k-trace zero for some algebraically closed subfield $k \subset F$, let $X \subset A$ be a subvariety, $\gamma \subset A$ a finitely generated subgroup and

$$\gamma_{div} = \{x \in A;\ \exists\ n \text{ such that } nx \in \gamma\}$$

Then there exist abelian subvarieties $B_1,\dots, B_m \subset A$ and points $a_1,\dots, a_m \in A$ such that

$$\gamma_{div} \cap X \subset \bigcup_{i=1}^{m} (B_i + a_i) \subset X$$

The above statement without the trace condition was conjectured by Serge Lang; recently G. Galtings proved a weaker form S. Lang's conjecture in which γ_{div} is replaced by γ.

Note also that in $[B_7]$ we proved an "infinitesimal analogue" of (B.1).

APPENDIX C. FINAL REMARKS AND QUESTIONS

We start with some intriguing short questions:

(C.1) Does there exist non-affine Δ_o-groups? (compare with (V. 2.14)). We think there should be; a candidate would perhaps be $E(A)_\Delta$ where A is an elliptic curve over $\mathcal{U}$ which is not defined over $\mathcal{K}$.

(C.2) Let G be an irreducible algebraic D-group (D = K[P], char K = 0). Is the linear part of G defined over K^D? This is crucial for understanding P(G) for arbitrary (non-linear, non-commutative) G.

(C.3) Let G be an affine algebraic D-group (D = K[P], char K = 0). Are the weights of G necessarily P-constants? (see (II. 3.12) for background). This is quite significant since we dispose of a very explicit description of affine algebraic D-groups all of whose weights are P-constant, cf. (II. 3.16), (V. 4.3).

(C.4) A hard task is to extend our theory from Chapters 1-4 by shifting from Spec K to a more general k-scheme, in particular a k-variety S ($k = \mathbf{C}$) and build a bridge between our theory and Deligne's theory of linear differential equations with regular singular points [Del_2]. Vector bundles on S with connections are in fact group schemes over S (with fibres vector groups) on which the algebra of differential operator D_S on S acts compatibly with multiplication, inverse and unit. So the theory we expect is an analogue of Deligne's in which algebraic vector groups G_a^N are replaced by more general algebraic groups.

More precisely, Deligne's theory in [Del_2] shows in particular that there is an equivalence between the "analytically defined" category

$(*)_{top}$ {local systems of vector spaces on S^{an}}

where S is a fixed smooth algebraic $\mathbf{C}$ - variety and the "algebraically defined" category

$(*)_{alg}$ {regular D_S - modules on S}

(here vector spaces are assumed finite dimensional and D_S-modules are assumed $\mathcal{O}_S$-coherent hence locally free).

One is then tempted to inspect relations between the "analytically defined" category:

$(**)_{top}$ {local systems of Lie groups on S^{an}}

and the "algebraically defined" category:

$(**)_{alg}$ {"regular" algebraic D_S-groups}

Here an algebraic D_S-group should be of course a group scheme over S on which D_S acts compatibly with multiplication, inverse and unit, while "regular" should mean at least that the D_S-module of relative Lie algebras is regular in Deligne's sense (compare with our regularity criterion (III. 3.9)).

Note that local systems of Lie groups were studied in [Ha]. We pose here the problem of their algebraisation. It is natural to start with the simplest examples of Lie groups hich are not vector spaces. Let's take for instance local systems on S^{an} whose fibres are $(\mathbf{C}^*)^M$ (these are equivalent to representations of $\Pi_1(S^{an})$ in $GL_M(\mathbf{Z})$). What local systems of this type come from algebraic D_S-groups? Of course, group shemes which become tori after passing to a finite etale covering of S lead to finite monodromy groups. But at least for even $M = 2g$ one can get infinite monodromy groups by taking non-trivial abelian schemes $f: A \to S$ of relative dimension g and then consider their extension $E(A) \to S$ by $(R^1 f_* \mathcal{O}_A)^v$; there is natural structure of algebraic D_S-group on E(A)/S whose monodromy is that of $R^1 f_*^{an} \mathbf{Z}$. Of course the fibres of E(A)/S are analytically isomorphic to $(\mathbf{C}^*)^M$. There are however topological restrictions for $R^1 f_*^{an} \mathbf{Z}$ (e.g. if S is a curve the latter should be quasi-unipotent at infinity). This shows that "algebraisation" should not be always possible. It is certainly not unique either. So the taks is to replace $(**)_{top}$ and $(**)_{alg}$ by some other related categories which should be equivalent to eachother and then to develop a dictionary between the topological and the algebraic setting.

(C.5) In $[B_8]$ we proved the following theorem: for any irreducible algebraic $\mathcal{U}$-group G there is a morphism of Δ-groups $\hat{G} \to \widehat{GL}_N$ (for some $N \geq 1$) whose kernel is a (commutative) Δ_0-group with no non-trivial linear representation. This shows how the study of Δ-groups appearing as Δ-subgroups of $\hat{G}$ for some algebraic $\mathcal{U}$-group G "reduces" to the study of linear Δ-groups and of Δ_0-groups with no non-trivial linear representations. So we are provided with a strong aposteriori motivation for our study of Δ_0-groups with no non-trivial linear representation!

REFERENCES

[BBM] **P. Berthelot, L. Breen, W. Messing,** Théorie de Dieudonné cristalline II, Lecture Notes in Math. 930, Springer 1982.

[BS] **A. Borel, J.P. Serre,** Théorèmes de finitude en cohomologie galoisienne, Comment. Math. Helvetici, 39(1964), 111-164.

$[B_1]$ **A. Buium,** Differential Function Fields and Moduli of Algebraic Varieties, Lecture Notes in Math. 1226, Springer 1986.

$[B_2]$ **A. Buium,** Ritt schemes and torsion theory, Pacific J. Math., 98(1982), 281-293.

$[B_3]$ **A. Buium,** Splitting differential algebraic groups, J. Algebra, 130, 1(1990), 97-105.

$[B_4]$ **A. Buium,** Birational moduli and non-abelian cohomology, I, Compositio Math. 68(1988), 175-202.

$[B_5]$ **A. Buium,** Birational moduli and non-abelian cohomology, II, Compositio Math. 71(1989), 247-263.

$[B_6]$ **A. Buium,** Intersections in jet spaces and a conjecture of S. Lang, to appear in Annals of Math.

$[B_7]$ **A. Buium,** Infinitesimal study of differential polynomial functions, Preprint INCREST 1990.

[B_8] **A. Buium,** Geometry of differential polynomial functions, Preprint INCREST 1990.

[C_1] **P. Cassidy,** Differential algebraic groups, Amer. J. Math. 94(1972), 891-954.

[C_2] **P. Cassidy,** Unipotent differential algebraic groups, in Contributions to Algebra, Academic Pres, New York 1977.

[C_3] **P. Cassidy,** The differential rational representation algebra of a linear differential algebraic group, J. Algebra 37(1975), 223-238.

[C_4] **P. Cassidy,** The classification of the semisimple differential algebraic groups and the linear semisimple differential algebraic Lie algebras, J. Algebra

[Del_1] **P. Deligne,** Théorème de Lefschetz et critères de dégénerescence de suites spectrales, Publ. Math. IHES 95(1968), 107-126.

[Del_2] **P. Deligne,** Equations Differentielles à Points Singuliers Réguliers, Lecture Notes in Math. 163, Springer 1970.

[D] **M. Demazure,** Schémas en groupes réductifs, Bull. Soc. Math. France, 93(1965), 369-413.

[DG] **M. Demazure, P. Gabriel,** Groupes Algébriques, North Holland, 1970.

[G] **A. Grothendieck,** Fondements de la Géometrie Algébrique, Sem. Bourbaki 1957-1962.

[GS] **R. Gerard, A. Sec,** Feuilletages de Painlevé, Bull. Soc. Math. France, 100(1972), 47-72.

[GD] **A. Grothendieck, M. Demazure,** Schémas en groupes I, Lecture Notes in Math. 151, Springer 1970.

[Ha] **H. Hamm,** Differential equations and local systems, Conference at Steklov Institute in Moscow, March 1989.

[Har] **R. Hartshorne,** Algebraic Geometry, Springer Verlag, 1977.

[H] **G. Hochschild,** Basic Theory of Algebraic Groups and Lie Algebras, Springer 1981.

[HM_1] **G. Hochschild, D. Mostow,** On the algebra of representative functions of an analytic group, Amer. J. Math. 93(1961), 111-136.

[HM_2] **G. Hochschild, D. Mostow,** Analytic and rational automorphisms of complex algebraic groups, J. Algebra 25, 1(1973), 146-152.

[J] **J. Johnson,** Prolongations of integral domains, J. Algebra 94(1) 1985, 173-211.

[J] **J.P. Jouanolou,** Equations de Pfaff Algébriques, Lecture Notes in Math. 708, Springer 1979.

[Ka] **N. Katz,** Nilpotent connections and the monodromy theorem; applications of a result of Turritin, Publ. Math. IHES 39(1970), 175-232.

[KO] **N. Katz, T. Oda,** On the differentiation of the de Rham cohomology classes with respect to parameters, J. Kyoto Univ. 8(1968), 119-213.

[Kei] **W. Keigher,** Adjunction and commonads in differential algebra, Pacific J. Math., 59(1975), 99-112.

[K_1] **E. Kolchin,** Differential Algebra and Algebraic Groups, Academic Press, New York 1973.

[K_2] **E. Kolchin,** Differential Algebraic Groups, Academic Press, New York 1985.

[K$_3$] **E. Kolchin,** Constrained extensions of differential fields, Adv. in Math. 12(1974), 141-170.

[Ma] **J.I. Manin,** Rational points of algebraic curves over function fields, Izvestija Akad. Nauk SSR, Mat. ser. t. 27(1963), 1395-1440.

[M] **M. Matsuda,** First order algebraic differential equations, Lecture Notes in Math. 804, Springer 1980.

[MO] **H. Matsumura, F. Oort,** Representability of group functors and automorphisms of algebraic schemes, Invent. Math. 4(1967), 1-25.

[MM] **B. Mazur, W. Messing,** Universal Extensions and One Dimensional Crystalline Cohomology, Lecture Notes in Math. 370, Springer 1974.

[Mu] **D. Mumford,** Geometric Invariant Theory, Springer 1965.

[N] **K. Nishioka,** A note on the transcendency of Painlevé's first transcendent, Nagoya Math. J. 109(1988), 63-67.

[NW] **W. Nichols, B. Weisfeiler,** Differential formal groups of J.F. Ritt, Amer. J. Math. 104(5) 1982, 943-1005.

[Li] **D. Lieberman,** Compactness of the Chow scheme, Sem. Norguet 1976, Lecture Notes in Math. 670, Springer 1978.

[R] **J.F. Ritt,** Differential Algebra, Amer. Math. Sci. Colloq. Publ. 33(1950).

[Ro] **M. Rosenlicht,** Some basic theorems on algebraic groups, Amer. J. Math. 78(1956), 401-463.

[Se] **J.P. Serre,** Groupes Algébriques et Corps de Classes, Hermann, Paris 1959.

[Sei] **A. Seidenberg,** Abstract differential algebra and the analytic case I, II, Proc. AMS 9(1958), 159-164 and 23(1968), 689-691.

[Sw] **M. Sweedler,** Hopf Algebras, Benjamin, New York 1969.

[T] **M. Takeuchi,** A Hopf theoretic approach to the Picard Vessiot theory, J. Algebra.

[U] **H. Umemura,** Sur les sous-groupes algebriques primitifs du groupe de Cremona a 3 variables, Nagoya Math. J. 79(1980), 47-67.

[V] **E. Viehweg,** Weak positivity and additivity of the Kodaira dimension II: the Torelli map, in : Classification of Algebraic and Analytic Manifolds, Birkhauser 1983, 567-584.

INDEX OF TERMINOLOGY

INDEX OF NOTATIONS